Thermodynamics of Complex Systems

Principles and applications

Thermodynamics of Complex Systems

Principles and applications

Vladimir N Pokrovskii
Institute of Chemical Physics, Russian Academy of Sciences, Moscow, Russia

IOP Publishing, Bristol, UK

ISBN 978-0-7503-3451-8 (ebook)
ISBN 978-0-7503-3449-5 (print)
ISBN 978-0-7503-3452-5 (myPrint)
ISBN 978-0-7503-3450-1 (mobi)

DOI 10.1088/978-0-7503-3451-8

Version: 20200801

IOP ebooks

British Library Cataloguing-in-Publication Data: A catalogue record for this book is available from the British Library.

Published by IOP Publishing, wholly owned by The Institute of Physics, London

IOP Publishing, Temple Circus, Temple Way, Bristol, BS1 6HG, UK

US Office: IOP Publishing, Inc., 190 North Independence Mall West, Suite 601, Philadelphia, PA 19106, USA

Contents

Preface xi

Author biography xiii

Part I Principles of non-equilibrium thermodynamics

1 The concept of a thermodynamic system 1-1

1.1 Composition of a thermodynamic system 1-1
1.2 Fundamental variables and work 1-3
1.3 Entropy and absolute temperature 1-4
1.4 The fluxes and thermodynamic forces 1-6
1.4.1 Fluxes of heat 1-6
1.4.2 Fluxes of substances 1-7
1.5 Non-equilibrium thermodynamics 1-9
1.5.1 Internal variables 1-9
1.5.2 The problem of formulation 1-10
References 1-11

2 The characteristic functions in non-equilibrium 2-1

2.1 The change of thermal energy 2-1
2.2 The fundamental principles 2-3
2.2.1 Law of conversion and conservation of energy 2-3
2.2.2 The arrow of time 2-4
2.2.3 Entropy and temperature 2-4
2.3 Non-equilibrium entropy 2-6
2.3.1 Entropy as a function of internal variables 2-6
2.3.2 Two components of entropy 2-7
2.4 Potential functions 2-9
2.4.1 Total internal energy 2-9
2.4.2 Free energy 2-10
2.4.3 Enthalpy 2-10
2.4.4 Gibbs free energy 2-11
2.5 Entropy as a function of state 2-11

2.6	Chemical potential	2-12
2.7	The peculiarities of thermodynamic functions	2-13
	References	2-14
3	**Dynamics of internal variables**	**3-1**
3.1	Internal variables of complexity	3-1
3.2	Preferable values of internal variables	3-2
3.3	Evolution of internal variables	3-4
	References	3-5
4	**Non-equilibrium steady states**	**4-1**
4.1	Dynamics of internal variables	4-1
4.2	Entropy near a steady state	4-2
4.3	Production of entropy	4-4
4.4	On the criterion of stability of stationary states	4-5
4.5	Relationship between fluxes and thermodynamic forces	4-5
	4.5.1 Symmetry of kinetic coefficients	4-6
4.6	The simplest samples	4-7
	4.6.1 Heat flux	4-7
	4.6.2 Diffusion	4-9
	4.6.3 The couple fluxes	4-10
	References	4-12
5	**Statistical interpretation of thermodynamic characteristics**	**5-1**
5.1	Two approaches to the thermodynamic system	5-1
	5.1.1 Macroscopic description	5-1
	5.1.2 Microscopic description	5-2
	5.1.3 How to compare two approaches?	5-2
5.2	Introduction of statistical ensemble	5-3
5.3	Statistics of an isolated system	5-6
	5.3.1 Microcanonic distribution	5-6
	5.3.2 Entropy of an isolated system	5-6
5.4	Statistics of closed system	5-7
	5.4.1 Canonical distribution	5-8
	5.4.2 Thermodynamic characteristics	5-10

5.5 The principle of entropy maximum 5-11
5.5.1 Classical functional 5-11
5.5.2 Unconventional functionals 5-15
5.6 Concluding remarks 5-15
References 5-16

Part II From molecules to the living organism

6 Chemical transformations 6-1

6.1 Kinetics of chemical transformations 6-1
6.1.1 Description of reacting mixtures 6-1
6.1.2 Examples of elementary reactions 6-3
6.2 Thermodynamics of chemical transformations 6-6
6.2.1 Thermodynamic potential and affinity 6-7
6.2.2 Entropy of the reacting mixture 6-8
6.2.3 Example of non-equilibrium steady state 6-9
6.3 Brusselator 6-11
6.3.1 A chemical clock 6-11
6.3.2 Kinetics of brusselator 6-13
6.3.3 Thermodynamics of brusselator 6-16
6.4 The problem of self-organization 6-19
References 6-20

7 Dynamics of complex fluids 7-1

7.1 The equation of continuity and the law of conservation of momentum 7-1
7.2 The law of conservation of energy and balance of entropy 7-2
7.3 Thermodynamic fluxes and relaxation processes 7-5
7.4 The principle of relativity for slow motions 7-6
7.5 Viscoelastic fluids 7-8
7.6 Different forms of constitutive relation 7-10
References 7-12

8 Creation of complexity 8-1

8.1 Complexity and information 8-1
8.1.1 Index of complexity 8-1
8.1.2 Concept of information 8-2

8.2 Statistical interpretation of ideal gas 8-3
8.2.1 The phase space 8-3
8.2.2 The number of microstates of the system 8-4
8.2.3 Thermodynamic probability 8-6
8.2.4 Entropy and temperature 8-6
8.3 Does a thermodynamic system contain information? 8-8
8.4 Production of complexity 8-9
8.4.1 Elementary production cycle 8-10
8.4.2 Production of commodities 8-12
References 8-15

9 Problems of thermodynamics of living organisms 9-1

9.1 The living organism as a thermodynamic system 9-1
9.1.1 Organism as an ensemble of chemical reactions 9-2
9.1.2 Hierarchy of internal variables 9-3
9.1.3 Complexity and entropy 9-4
9.2 The growth and development of the living organism 9-5
9.2.1 The growth equation 9-6
9.2.2 The testing of the growth equation 9-8
9.3 Some peculiarities of the living organism 9-10
9.3.1 The control and regulate system 9-11
9.3.2 Mechanism of mobility 9-12
9.4 Evolution of living matter 9-13
9.4.1 Ontogenesis and phylogenesis 9-13
9.4.2 Where is the beginning? 9-14
References 9-14

Part III Approaching the superorganisms

10 Dynamics of populations 10-1

10.1 Dynamics of a separate population 10-1
10.1.1 Balance equation 10-1
10.1.2 Exponential growth 10-2
10.1.3 The logistic growth 10-2
10.1.4 Migration 10-4
10.1.5 A wave in logistic population 10-6
10.1.6 The composition of a population 10-7

10.2 Interacting populations 10-8
10.2.1 The model of competing populations 10-8
10.2.2 The predator–prey model 10-10
10.2.3 Dynamics and thermodynamics of biogeocoenosis 10-12
10.3 Peculiarities of development of human population 10-14
10.3.1 Empirics of population growth 10-14
10.3.2 The laws of growth of a human population 10-16
10.3.3 Thermodynamics of the human population 10-18
References 10-20

11 Problems of sociodynamics 11-1

11.1 Architecture of social organization 11-1
11.2 Functional social structure 11-3
11.2.1 The status of an individual and socioconfiguration 11-3
11.2.2 Strata, domains, classes and castes 11-5
11.2.3 Groups and organizations 11-7
11.3 Dynamics of the social structure 11-8
11.3.1 Mechanism of changes of the system of values 11-8
11.3.2 Mechanism of changes of the functional structure 11-10
11.3.3 Kinetic equation 11-11
11.3.4 Rules of change of individual status 11-12
11.3.5 Mechanism of decision-making 11-13
11.3.6 Formation of collective opinion 11-15
11.4 Institute of economic activity 11-16
11.4.1 The scheme of production process 11-17
11.4.2 Problem of distribution of the social product 11-19
11.4.3 Stratification on the wealth 11-23
References 11-29

12 Dynamics of cognition 12-1

12.1 Imperative to putting in order 12-1
12.1.1 Level of unconscious mind 12-2
12.1.2 Level of consciousness 12-2
12.1.3 Statements we believe in 12-3
12.1.4 The events 12-4

12.2 Description of the events 12-5
12.2.1 Space and time 12-5
12.2.2 Principle of self-preservation 12-6
12.2.3 The picture of our world 12-7
12.3 Mathematical modeling 12-8
12.3.1 Description tools 12-9
12.3.2 Scheme of the process of cognition 12-9
12.3.3 The system of physics models 12-12
12.4 Science as a social institution 12-14
12.4.1 Social function of science 12-14
12.4.2 About scientific method 12-15
12.4.3 Organization of scientific research 12-16
12.4.4 A career in science 12-16
References 12-18

Preface

The reader is invited to look at the complex systems that we encounter in reality (living biological organisms and populations of species, including the human population with special subsystems: a national economy, monetary system, science etc) from the point of view of thermodynamics. A complex system can be considered as a thermodynamic system, which is described not only by *fundamental variables*, characterizing thermodynamic equilibrium, but also by a set of extra, *internal variables* that describe deviations of the thermodynamic system from the equilibrium situation. The existence of internal variables is the distinguishing mark that separates complex thermodynamic systems from simple ones. In fact, a very complex system appears to be a metastable construction, described by internal variables; the conventional thermodynamics variables like volume, pressure and so on do not play an essential role in description of behaviour of a very complex system. However, energy and entropy appear to remain the fundamental concepts in thermodynamics of complex systems.

Non-equilibrium thermodynamics appears to be a universal basis for the description of behavior of a complex system, and, in the first chapters of the book, I try to represent the main concepts and principles of non-equilibrium thermodynamics in a way applicable both to open systems and to situations far from equilibrium. The consecutive use of the concept of *thermodynamic internal variables* allows us to take into account the complexity of the system and to describe its dynamics. The exposition begins with the definition of a thermodynamic system and is developed before the reader step by step; although, perhaps, it could be easier for him to have preliminary knowledge of some notions from physics and thermodynamics. In the subsequent chapters, some samples of application of the general theory of thermodynamics of complex systems is considered. I hope that the presented concise formulation can be helpful for those who are occupied with applications of thermodynamics to the description of real objects.

The scheme of disposition was born when I was delivering lectures to my students of applied mathematics in the framework of the Course on Methods of Mathematical Modeling many, many years ago. So, I suppose, the book can be used by the students of different specialties who are going to develop mathematical models of real phenomena. The form of lectures did not assume long excursions into the history of concepts and principles of thermodynamics, so that my list of literature is not as large as it should be. I am realizing that some important work and authors may not be mentioned (I am sorry for that!), and this is only the consequence of the situation.

The drafts of the chapters of the book were presented for discussion on the site ResearchGate (https://www.researchgate.net/). I am grateful to those who sent me comments on my work: Ben Akih kumgeh, Théodore Bouchez, Erik Brown, Delton

Chen, Vera Maura Fernandes de Lima, Marc Le Goc, Purushottam D Gujrati, Luciano Martinez-Balbuena, Wolfgang Muschik, Alfredo Pereira Junior, Marko Popovic, Vit Průša, Bartolomé Sabater, Robert Shour, A M Toikka, Mohammad Wahiduzzaman. Their suggestions have helped me, I hope, to improve the text. Nevertherless, all possible mistakes are mine. I shall be glad to hear any comments on the book.

Vladimir N Pokrovskii
Moscow
http://www.ecodynamics.narod.ru
May 2020

Author biography

Vladimir N Pokrovskii

Vladimir Pokrovskii is a Professor of Applied Mathematics. He is an expert in mathematical modelling of complex natural and artificial systems. He graduated from the Department of Theoretical Physics, Tomsk State University. His qualification was acquired at the Institute of Chemical Physics of AH USSR, investigating the behaviour of suspensions and polymers. He is known as a prominent researcher of dynamics of polymers: the second edition of his monograph *The Mesoscopic Theory of Polymer Dynamics*, issued by Springer in 2010, is included in the list of Best Reference Books on Polymer Dynamics, used by students of top universities, institutes and colleges in the USA. He has developed an original course of lectures *Methods of mathematical modelling*, delivered to students of applied mathematics of the Altai Polytechnic Institute and the Moscow Institute of Economics and Statistics, that includes classical examples from areas of mechanics, physics, chemistry, biology and economy. Persistent attempts to understand the essence of economic processes have provided material for the book *Econodynamics. The Theory of Social Production*, the third edition of which was published by Springer in 2018. Vladimir Pokrovskii has authored some other monographs (published in Russian) and more than 150 papers (in Russian and English).

Part I

Principles of non-equilibrium thermodynamics

IOP Publishing

Chapter 1

The concept of a thermodynamic system

To define *a thermodynamic system* means, first of all, to allocate some portion of the substance or matter, on which our attention is focused. This portion should not be too small; we assume also that the system consists of many separate cooperating particles (atoms and molecules), moving in a certain field. The laws of movement of the particle can be known, though we do not need to know them to formulate the macroscopic principles; moreover, a detailed microscopic description of the system appears to be possible in principle only. To describe gross characteristics, or large scale properties of the system, one has to introduce a set of new variables. To provide a macroscopic description, the thermodynamics solves, first of all, the problem of development of the methods of the description of behaviour of the assembly of many particles, as a whole system under various influences.

1.1 Composition of a thermodynamic system

The thermodynamic system is considered as an assembly of molecules of various chemical substances, the quantity and some characteristics of which sometimes (but not always) are known. The evidence of complex composition of macroscopic bodies was observed during many years[1]. We assume that the numbers of molecules (numbers of moles) of various substances in the considered system can be set

$$n_1, n_2, \dots, n_k. \tag{1.1}$$

[1] For an illustration it is possible to consider some volume of water. Since ancient times it is known that properties of water do not change if the given volume is divided into two parts. Anything from these two volumes can be also divided into two parts, while water remains water. We can recognize properties of water in each portion, but a question arises: how long is it possible to continue the process of division? Philosophers of ancient Greece believed that division cannot last indefinitely long. Finally, one can reach an indivisible particle named an atom ($\alpha\tau o\mu o\sigma$—not capable to be cut, indivisible), up to the last portion of volume, which still possesses the properties of water. Now, after many centuries, we know that a glass of water, for example, contains approximately 6×10^{24} molecules of water H_2O and this quantity (approximately 180 g) can be divided in two only 82 or, possibly, 83 times to reach the limit of division.

doi:10.1088/978-0-7503-3451-8ch1

One has an equivalent description, when the concentrations of substances are given

$$c_1 = \frac{n_1}{n},\ c_2 = \frac{n_2}{n},\ \dots,\ c_k = \frac{n_k}{n}, \qquad n = \sum_{\alpha=1}^{k} n_\alpha,$$

where n is a total of molecules in the system[2]. Considering the thermodynamic system, the complete chemical composition of the system appears to be unknown; one is interested in the description of changes of composition, which are occurring due to internal chemical reactions or fluxes of substances from the environment of the system.

As a system, obeying the laws of classical or quantum mechanics[3], the assembly of particles is characterized by a certain quantity of *energy*, one speaks about *internal energy U*, assuming the system in the rest, and this quantity remains one of the main characteristic of thermodynamic system. The value of internal energy depends on the composition of the system and changes under external influences on the system. The change of composition provokes the change of the internal energy of the system that can be written as

$$\Delta U = \sum_{\alpha=1}^{k} h_\alpha\, \Delta n_\alpha. \tag{1.2}$$

The quantity h_α presents the change of internal energy that is occurring in the system with coming of a particle (mole) of substance. The quantity is introduced by equation (1.2) as intensive quantity, so that it can depend on the ratios of extensive quantities, that is on the concentrations $c_\alpha = n_\alpha / n$

$$h_\alpha = h_\alpha(c_1, c_2, \dots, c_n). \tag{1.3}$$

The energy, obtained by the system at variation of composition, depends, apparently, on the conditions at which the process occurs and, consequently, the specific enthalpy depends also, except for concentration, on other characteristics of the thermodynamic system, which will be discussed in the subsequent sections (see section 2.9).

Note that the change of number of particles in the system can occur due to both internal transformations and external fluxes of substance. A system, in which there are external fluxes of substances, is considered an *open system*. The numbers of particles in fluxes in and out the system ought to be specified. Otherwise, if there are

---

[2] One can use also the mass concentration, defined by equation

$$c_\alpha = \frac{m_\alpha}{V} n_\alpha.$$

[3] The microscopic specification of a system is used for the introduction of a statistical description of the thermodynamic system. In greater detail, it is possible to see, for example, the fifth chapter (section 5.1.2) for classic mechanics.

no fluxes of substances at all, the system is regarded as *closed* or even *isolated.* We shall return to the classification later.

1.2 Fundamental variables and work

We are concentrating on a collection of matter, which is considered as a *thermodynamic system* and must be specified further by introducing some characteristics of the state of the system. Everything external to the selected portion of matter is called *the surroundings* or the environment (thermostat), so that certain *borders* of the thermodynamics system exist. If we consider the system of stable composition, the borders are non-penetrable for matter and there are no fluxes of substances into or out of the system. The borders are fixed by a set of variables, and for the certain thermodynamic system the set of the parameters allocating the system from the environment is usually known from experiments. In the following we shall name such parameters as *fundamental variables* and designate $x_1, x_2, \dots, x_n$. The specified variables allocate the thermodynamic system, defining its borders. In the simplest case, the set of fundamental variables reduces to the only variable—to volume V, containing the certain quantity of substance in a gas or liquid phase.

The interaction between the system and its surroundings is considered to be performed via work of the fundamental variables $x_1, x_2, \dots, x_n$, and therefore is a form of energy in transit. An expression for work of the system has to be written as

$$\Delta A = \sum_{i=1}^{n} X_i \, \Delta x_i \tag{1.4}$$

where X_i, $i = 1, 2, \dots, n$ are thermodynamic forces which depend on the process of crossover from one state to another. The expression (1.4) will be everywhere used to represent work done by the system. The work done on the system has the opposite sign, however, it is not equal, in the general case, to the work done by the system. As a convention of sign, the quantity (1.4) is considered positive, while work done on the system is considered negative. The quantity ΔA stands for a small part of work. In the simplest case, when a system is characterized by the only fundamental variable—by volume V, work of the system is presented in the form

$$\Delta A = p \, \Delta V$$

where V is a volume of the system, p is internal pressure within the system, which need not be equal to the external pressure.

The work fulfilled by the system leads to variation of the internal energy of the system. In the simple case, when the reduction of energy is equal to the work, $\Delta U = -\Delta A$, the thermodynamic system can be considered as an *isolated* system, though usually it is not allowed work for an isolated system. However, there are cases, when $\Delta U \neq -\Delta A$. The inequality is connected with the possibility of exchange of energy between the system and environment in the form of thermal energy. In this case, the system is named a *closed* one, assuming the composition of the system is unchanged. The closed system has, generally speaking, no certain value of energy. The clarification of the role of thermal interaction was the basic difficulty, and the proper solution (Clausius 1865) has led to the modern formulation of thermodynamics.

1.3 Entropy and absolute temperature

The energy passed from one system to another because of the temperature distinction between the two systems is named *heat energy*, which can be identified only at crossing of the border of the system and, consequently, it is a form of energy in transit. As an agreement of a sign, the quantity of heat received by a system ΔQ is considered positive, while the quantity of heat given by the system, is considered negative. If borders of the closed system do not allow exchange of heat between the system and the environment, the closed system is named *isolated.*

The change of internal energy U is provoked by the exchange of energy between the system and the surroundings[4]. The internal energy of the system U is increasing due to the flux of heat coming into the system ΔQ and decreasing due to the work of the system ΔA; the energy coming with the fluxes of substances (1.2) is considered as a positive contribution, so that the change of the energy of the system can be written as

$$\Delta U = \Delta Q - \sum_{j} X_j \, \Delta x_j + \sum_{\alpha} h_\alpha \, \Delta n_\alpha. \tag{1.5}$$

This expression represents the standard form of the first law of thermodynamics, as it is stated in any text-book on thermodynamics. All quantities in equation (1.5) are characteristics of the thermodynamic system (not the environment), and the written relation is valid for any processes (reversible or irreversible) and situations.

There appears a difficulty, when one intends to define internal energy as a function of variables of a state. The matter is that the heat Q cannot be an argument of a function of state, this quantity is not a characteristic of the system, it is not a property of the system, it is heat obtained by the system. The designation ΔQ stands for a small part of heat, while we use the symbol Δ to show a change of the amount of a quantity which is not a function of system.

In the earlier days of development of the science of heat, it was supposed, that the stream of heat is connected with a stream of a special substance—phlogiston that flows in and out the thermodynamic system. Phlogiston had not been detected, and the problem, as it is well known, was solved by Clausius (1865) by postulating the existence of a such quantity S, called entropy and defined as

---

[4] It was experimentally shown that energy is a measure of equivalence of the various macroscopic phenomena. The steam engines of Newcomen (from 1712) and Watt (from 1774) demonstrated that heat performs mechanical work. On the other side, work can produce heat. The historically most important experiment illustrating the reverse transformation of work into heat was performed in Munich by Count Rumford (1798): he caused water to boil, when a cannon barrel was being bored. In 1843 through his famous experiments Joule was led to the postulate that heat and work were equivalent in some sense. Mechanical work and heat are interchangeable, i.e. heat may be converted into mechanical work as in a steam engine, or mechanical work can be converted into heat which occurs as the barrel of a cannon is bored.

$$d\,S = \frac{\Delta Q}{T},$$

where T is an absolute temperature that will be discussed later.

Generalizing the concept of entropy for the systems with variable composition, one can write the total differentials of the total internal energy and entropy (in a reversible process) as

$$dU = T\,dS - \sum_j X_j\,dx_j + \sum_\alpha \mu_\alpha\,\Delta n_\alpha, \tag{1.6}$$

$$dS = \frac{\Delta E}{T}, \quad \Delta E = \Delta Q + \sum_\alpha \eta_\alpha\,\Delta n_\alpha. \tag{1.7}$$

When determining the entropy S, we are forced to divide the total contribution of chemical energy h_α, brought to the system by particles of substances, into two parts $h_\alpha = \eta_\alpha + \mu_\alpha$. One part with the chemical potential μ_α, contributes to the potential internal energy, as defined by Gibbs (1875); the other part with the specific value η_α is converted immediately into the thermal energy of the system[5]; it is convenient to introduce in the expression (1.7) the symbol E for internal thermal energy of the system.

In contrast to equation (1.5), which is valid for any situations, equations (1.6) and (1.7) are valid for the specific processes, when there is no need to introduce extra variables, apart from fundamental ones, so-called reversible processes. Note that, while equation (1.6) defines internal total energy as a function of variables S, $\boldsymbol{x}$ and $\boldsymbol{n}$, equation (1.7) does not define entropy as a function[6]; arguments of the function of entropy will be discussed later (see section 2.5).

The notion of entropy, as a measure of internal heat energy, was firmly integrated by science, and the attempts to find phlogiston were abandoned. Simultaneously with entropy S, the absolute temperature T with an absolute thermodynamic scale of temperature (see any text-book on thermodynamics) is defined. The absolute temperature T is always positive and has a constant reference point. The temperature can be interpreted in terms of other physical quantities and is measurable, or,

[5] The definition of the quantities will be done in section 2.8. The quantity μ_α is chemical potential and η_α is entropy contribution to thermal energy by a particle (mole) of the αth substance; due to equation (2.27), the sum $\eta_\alpha + \mu_\alpha$ denotes specific enthalpy of the substance. Researchers use, in line with concepts of chemical potential and partial enthalpy, a notion of 'partial molar entropy of the exchanged chemical species under the conditions of the system boundary' (von Stockar and Liu 1999) that is identical to the quantity η_α/T. Having in mind to develop the theory of open systems, which could be applied to biological objects, we cannot allow omitting the important quantity η_α.

[6] Some researchers (see, for example, Sommerfeld 1952, chapter 2, section 14) do define the function by the expression

$$TdS = dU + \sum_j X_j\,dx_j - \sum_\alpha \mu_\alpha\,\Delta n_\alpha.$$

One can note, taking equation (1.6) into account, that the written relation is not a definition of function, but an uncertain differential equation for quantity S. Usage of this relation brings some inconsistency.

one can say, that it has physical meaning and empirical justification for its existence, as an intensive characteristic of the system. The temperature is usually measured in degrees, but, according to the connection with other physical quantities, it is energy. To omit the factor of transition from degrees to energy, it is convenient to agree that the temperature is measured in power units, as was proposed by Landau and Lifshitz (1986, p 52), which is accepted further at the writing of all equations. If the temperature is measured in degrees, to recalculate to power units, the factor of transition k, named the Boltzmann constant, is used. It has value

$$k = 1.38 \times 10^{-16}\ \text{erg/ grad}=1.38 \times 10^{-23}\ \text{J/ grad}.$$

It is assumed that the values of *entropy* and *temperature* can be attributed to every thermodynamic system, as to the object consisting of many 'particles'. Let us notice, however, that these specific quantities have been introduced by equation (1.6) for the reversible processes, which means that the processes of transition from one state to another are being fulfilled so slowly[7], that in each point in time the state of the thermodynamic system is unambiguously defined by its unique set of fundamental variables, which means that the system remains in the state that is referred to as *an equilibrium state*. The equilibrium macrostate is unique, it is a certain idealization of the reality, similar to an idealized motion of a body with constant velocity in mechanics, and in this sense, it is a natural state of a thermodynamic system. The thermodynamic system cannot go out of the equilibrium macrostate without external influences (from the environment).

1.4 The fluxes and thermodynamic forces

The situation of equilibrium is a certain idealization of the reality; the real thermodynamic system is an object of various influences of the environment that causes variations of thermodynamic system as far as it is allowed by the borders of the system. In the most general case we should admit that borders of the system allow an exchange not only of heat, but also substances. In this general case, the system is named *open*; we recall, the system is named *closed* if the borders are impenetrable for substances, but allow transit of energy in the form of heat, and *isolated*, if there is no exchange of heat and substances.

1.4.1 Fluxes of heat

Let us consider a system consisting of two adjacent subsystems with various temperatures, T_1 and T_2, which are incorporated in a combine system considered isolated. When two systems at different temperatures are brought in thermal contact

[7] More exactly, the slow fashion means that all characteristic times of the processes are much greater than corresponding relaxation times, $t \gg \tau_i$, and all variables practically have their equilibrium values. Such slow processes, when a state of thermodynamic system is determined only by fundamental variables, are called *reversible*. Otherwise, some of the relaxation times appear to be greater than a characteristic time of the process, which is called in this case *irreversible*. During such a process the interior of the system is permuted, and one needs in some extra variables (called internal ones) to describe the macrostate of the system more completely.

with each other, the combined system appears to be in a non-equilibrium state. It is known from evidence that there is a flux of heat from the system with higher temperature to the system with lower temperature, which appears to diffuse the difference in temperatures. The total energy of the combined system is constant, so that in any case, the change of heat within system 1 due to the flux from system 2 is the same (with the reverse sign) as the change of heat within system 2 due to the flux from system 1

$$\Delta Q_1 = -\Delta Q_2.$$

The heat fluxes provoke the changes of entropy in the compartments that, according to definition (1.7), can be written as

$$\Delta S_1 = \frac{\Delta Q_1}{T_1}, \quad \Delta S_2 = -\frac{\Delta Q_1}{T_2}.$$

In contrast to heat, the change of entropy in one compartment does not correspond to the change of entropy in the other department

$$\Delta S_1 \neq -\Delta S_2$$

and the change of entropy of the combined system can be estimated as

$$\Delta S = \frac{T_2 - T_1}{T_1 T_2} \Delta Q_1. \tag{1.8}$$

The change of entropy in this process is always non-negative and the trend to equilibrium can be interpreted as the trend of entropy to the highest value.

The results of the heat fluxes are the changing of values of temperatures of either system, so that a single value of temperature is established through the combined system. Equality of temperatures is a condition for thermal equilibrium between two systems or between two parts of a single system.

The inequality of temperatures in different points of the system provokes a flux of heat from the point with higher value to the point with lower value of temperature, so that one can write a local relation, for example, in the simplest form

$$J_T = -\lambda \, \nabla T, \quad \lambda > 0, \tag{1.9}$$

where coefficient of proportionality is named coefficient of conductivity of temperature (conductivity of heat).

1.4.2 Fluxes of substances

In a similar way we shall consider the system consisting of two adjacent subsystems with different concentrations of substances. If subsystems are separated by an impenetrable-for-substances partition, no variations in the system and subsystems is observed. After removal of the partition, changes of concentration of substances in both subsystems occur, and, if the compound system as a whole is isolated, the total energy of the system and number of particles in it do not change, so that the following relation, based on (1.2), can be written

$$\sum_{\alpha} \left(h_\alpha(c^{(1)}) - h_\alpha(c^{(2)})\right) \Delta n_\alpha^{(1)} = 0. \tag{1.10}$$

In an initial state, $\Delta n_\alpha^{(1)} \neq 0$ and therefore $h_\alpha(c^{(1)}) - h_\alpha(c^{(2)}) = 0$. This means that the quantity h_α, which is called the specific heat content of the substance α and is represented as the sum of two quantities $h_\alpha = \mu_\alpha + \eta_\alpha$ (see section 1.3), does not depend on the concentration.

From the recorded parity it does not follow that there is any preferential distribution of concentration, or any fluxes of substances, when the partition is removed. Mechanics can tell nothing about variations of concentration in subsystems; it does not follow from the laws of mechanics that the flow of matter is directed from a system with a higher concentration to a system with a lower concentration; we accept this as an empirical fact, which ought to be explained. To do this, we refer to the two quantities introduced by the relations (1.6) and (1.7): potential energy and entropy, the changes of which for each of the subsystems are defined as

$$\Delta(U_1 - E_1) = \sum_{\alpha} \mu_\alpha(c^{(1)})\, \Delta n_\alpha^{(1)}, \quad \Delta(U_2 - E_2) = -\sum_{\alpha} \mu_\alpha(c^{(2)})\, \Delta n_\alpha^{(1)},$$

$$\Delta S_1 = \frac{1}{T}\sum_{\alpha} \eta_\alpha(c^{(1)})\, \Delta n_\alpha^{(1)}, \quad \Delta S_2 = -\frac{1}{T}\sum_{\alpha} \eta_\alpha(c^{(2)})\, \Delta n_\alpha^{(1)}.$$

We believe that both chemical potential μ_α and the specific entropy flow η_α/T depend on concentration, and the changes in the quantities in one compartment do not correspond to changes in these values in another compartment.

The changes in the potential energy and entropy of the combined system after a certain amount of substance $\Delta n_\alpha^{(1)}$ has moved from subsystem 2 to subsystem 1 can be estimated as

$$\begin{aligned} \Delta(U - E) &= \sum_{\alpha} \left(\mu_\alpha(c^{(1)}) - \mu_\alpha(c^{(2)})\right) \Delta n_\alpha^{(1)}, \\ \Delta S &= \frac{1}{T}\sum_{\alpha} \left(\eta_\alpha(c^{(1)}) - \eta_\alpha(c^{(2)})\right) \Delta n_\alpha^{(1)}. \end{aligned} \tag{1.11}$$

We believe that the potential energy of the entire system cannot spontaneously increase, it can only decrease, and therefore the entropy of the combined system does not decrease with alignment. We can see, as in the previous example with heat, that the tendency to equilibrium can be interpreted as the tendency of entropy to the highest value. It follows also from the above speculations that the specific entropy η_α/T decreases with increasing concentration. In this case, the chemical potential μ_α is an increasing function of the concentration of the substance.

If concentration of a substance in the environment differs from concentration of a substance in the open system, there are processes of alignment of concentration. There is a flux of substances through the border of the system $\boldsymbol{J}_c$; it is convenient to reduce the situation to local consideration: the inequality of concentration in adjacent points of the system leads to a stream of substance from a point with higher value to a point with smaller value of concentration, so that it is possible to record for the flux of substance

$$J_C = -D\nabla c, \quad D > 0, \tag{1.12}$$

where D is the coefficient of diffusion. The process of alignment of composition is running until concentrations of substances appear equal in all points of the compound system—one speaks of the system coming to equilibrium. Equality of concentration is a condition of equilibrium of two systems or parts of a united system. The specific thermodynamic characteristics in various parts of the system also appear equal.

1.5 Non-equilibrium thermodynamics

Macroscopic behaviour of the thermodynamic system does not reduce to any principles of mechanics, and a phenomenological theory, based on the specific thermodynamic principles, has been developing. It was understood that specific quantity—entropy—plays, in line with energy, a prominent role in description of macroscopic behavior of the system. Evidence of this kind is considered in section 1.4, which shows that for the description of non-equilibrium phenomena one needs concepts of thermodynamic forces and fluxes. To describe the state of a thermodynamic system that deviates from the equilibrium state, one needs to introduce special quantities that are called *internal variables*.

1.5.1 Internal variables

Among internal variables, as we can see in the above examples, one can find variables which determine the local deviations of concentrations and temperature from equilibrium ones

$$\Delta c_\alpha = c_\alpha - c_\alpha^0, \ \alpha = 1, 2, \dots, n, \quad \Delta T = T_1 - T_2, \tag{1.13}$$

and the gradients of concentrations and temperature

$$\nabla c_\alpha, \ \alpha = 1, 2, \dots, n; \quad \nabla T. \tag{1.14}$$

These quantities have various values at various points of the thermodynamic system and characterize the non-equilibrium state of the system. Apart from these quantities, many other internal variables can appear during processes of transition from one state to another. Generally, one needs in a set of internal variables $\xi_1, \xi_2, \dots$, to describe deviation of the state of thermodynamic system from the equilibrium one.

The internal variables emerge, when the thermodynamic system is undergone with influences from the environment. The rate of emergence is connected with the history of fluxes $\boldsymbol{J}$ into the system

$$\frac{d\xi_i}{dt} = \mathcal{B}_i \frac{\infty}{s = 0}[\boldsymbol{J}(t - s)], \quad i = 1, 2, \dots. \tag{1.15}$$

Having been allowed to be on their own, the internal variables are tending to their unique equilibrium values, so that the system is tending to its stable equilibrium state. The local law of disappearing can be written as the relaxation equation for each internal variable

$$\frac{d\xi_i}{dt} = -\frac{1}{\tau_i}\left(\xi_i - \xi_i^{(0)}\right), \quad i = 1, 2, \ldots, \tag{1.16}$$

where $\tau_i = \tau_i(T, x_1, x_2, \ldots, x_n)$ is a relaxation time of corresponding variables. It is convenient to consider the initial value ξ_i^0 is equal to zero. The tending of the internal variables to disappear means that the equilibrium state is stable[8].

To investigate the non-equilibrium situation, one has to consider the two simultaneous processes: the emerging and disappearing of the internal variables. The first one is connected with external influences, the disappearing is a process obeying the internal laws. In the situations that are close to equilibrium, the disappearing is described by equation (1.16). The dynamics of the emerging and relaxation of internal variables in an arbitrary situation is considered in chapter 3.

Let us note that, when some of the times of relaxation of internal variables are very big as compared with times of external processes, the corresponding internal variables appear to be constant, thus constituting a metastable structure of the system. In the general case, the internal variables in the system represent fading memory about past interactions.

1.5.2 The problem of formulation

The thermodynamic description of equilibrium situations is fulfilled perfectly. However, it cannot be said about non-equilibrium situations: according to authoritative experts on *non-equilibrium thermodynamics* (Maugin and Muschik 1994, p 219), to date 'there is no formal consensus on how to formulate a thermodynamics of non-equilibrium'. The conformation of this assessment can also be found in other works, for example, it is possible to read in the article (Mesquita *et al* 1999): '... the thermodynamics of irreversible processes has not yet achieved a phenomenological formulation and a methodology which can be considered satisfactory'.

To consider the complex system, the principles of non-equilibrium thermodynamics ought to be formulated first of all, and, further on, in this part, an attempt is made to present, following previous investigation by the author (Pokrovskii 2005, 2013), the consequent formulation of the theory, in which we use an extra set of variables, so-called internal variables, for the fixation of a non-equilibrium state of a thermodynamic system. It leads to a foundation and some justification of the known

[8] Note that the reversible process can now be defined more accurately. The process of transition of a thermodynamic system from one macrostate to another is called *reversible* if the relaxation times of all internal variables τ_i are much smaller than the characteristic times of the transition process t; in this case, all fundamental variables actually have equilibrium values, and the internal variables are zero values. Otherwise, some of the relaxation times are greater than the characteristic time of the process, which, in this case, is called *irreversible*. In this case, some of the internal variables are excited and must be taken into account implicitly or explicitly.

(see, for example, Kondepudi and Prigogine 1999) parities of non-equilibrium thermodynamics. Turning to the regular description of non-equilibrium behaviour of a thermodynamic system, we are keeping in the mind the open thermodynamic system that intensively exchanges heat and substances with the environment. We shall notice, that use of internal variables is a universal method of considering of non-equilibrium situations, it is not a special 'thermodynamics with internal variables'.

Perhaps, the first, who used internal variables for the consecutive formulation of thermodynamics, was Leontovich (1983)[9]. An independent development of the theory is presented in the work (Maugin and Muschik 1994) where authors understand the internal variables as only the complexity internal variable (see further chapter 3), and interpret the theory, as classical thermodynamics of irreversible processes with additional variables of state and an appropriating axiom of local balance. There is an alternative version of the theory in which not the internal variables, but fluxes are used for the description of non-equilibrium situations. Such a description named 'the expanded irreversible thermodynamics' (Jou *et al* 2001) appears equivalent to the description by means of the internal variables in stationary cases, so as in these cases there is an algebraic correspondence between streams and internal variables. In general case there is no such correspondence: it is possible to imagine the case when the fluxes are absent, but the system is in a non-equilibrium state with nonzero internal variables, and descriptions do not appear equivalent. The usage of fluxes as characteristics of the state cannot be proved in a conclusive way, and the version with the internal variables remains preferential.

References

Clausius R 1865 Ueber verschiedene für die Anvendung bequeme Formen der Hauptgleichungen der mechanischen Wärmetheorie *Ann. Phys. Chem.* **125** 353–400

Gibbs J W 1875 On the equilibrium of heterogeneous substances *Trans. Conn. Acad. Arts Sci.* **III** 108–248 343–524

Gibbs J W 1928 *Equilibrium of Non-Homogenous Substances, 1895–1898: Collected works* (New York: Longman Green)

Jou D, Casas-Vázquez J and Lebon G 2001 *Extended Irreversible Thermodynamics* 3rd edn (Berlin: Springer)

Kondepudi D and Prigogine I 1999 *Modern Thermodynamics: From Heat Engines to Dissipative Structures* (Chichester: Wiley)

Landau L D and Lifshitz E M 1986 *Statistical Physics* vol 1 3rd edn (Oxford: Pergamon)

Leontovich M A 1983 *Vvedenie v termodinamiku. Statisticheskaya fizika [Introduction to Thermodynamics. The Statistical Physics]* (Moskva: Nauka, Glavnaya redaktsiya fiziko-matematicheskoy literatury)

Maugin G A and Muschik W 1994 Thermodynamics with internal variables part I. General concepts *J. Non-Equil. Thermodyn.* **19** 217–49

[9] The first Russian edition of *Introduction to Thermodynamics*, based on his pre-war and war-time lectures, appeared, as a separate book, in 1950.

Mesquita M V, Vasconcellos A R and Luzzi R 1999 Irreversible processes in the context of a non-equilibrium statistical ensemble formalism *Phys. Scr.* **59** 257–65

Pokrovskii V N 2005 A derivation of the main relations of non-equilibrium thermodynamics *Eur. J. Phys.* **26** 769–81

Pokrovskii V N 2013 A derivation of the main relations of non-equilibrium thermodynamics *ISRN Thermodyn.* **2013** 906136

Sommerfeld A 1952 *Thermodynamik und Statistik—Vorlesungen uber theoretische Physik Band 5* ed F Bopp and J Meixner (Wiesbaden: Diederichsche Verlagsbuchhandlung) Translation: Sommerfeld A 1964 *Thermodynamics and Statistical Mechanics—Lectures on Theoretical Physics Volume V* ed F Bopp and J Meixner (New York: Academic) translated by J Kestin

von Stockar U and Liu J-S 1999 Does microbial life always feed on negative entropy? Thermodynamic analysis of microbial growth *Biochim. Biophys. Acta* **1412** 191–211

Chapter 2

The characteristic functions in non-equilibrium

The characteristic distinction of non-equilibrium situations from equilibrium ones is the presence of internal variables that describe the deviation of the system from the equilibrium. Though in some cases, the internal variables are not used explicitly, it is important to keep in mind their existence for correct description of behaviour of the thermodynamic system. In some cases, the internal variables and the processes connected with them are determining the essence of the thermodynamic system. In this chapter, to describe the role of internal variables in the processes of interaction with surroundings, we shall subsequently introduce the internal variables into the main definitions and relations of thermodynamics.

2.1 The change of thermal energy

The thermodynamic system (in equilibrium or not) is considered as a collection of particles, to which a certain amount of energy U is ascribed. We called this amount the *total internal energy*, which changes only due to the influences from the environment, which can be mechanical work and/or fluxes of heat and matter. We can consider also the internal thermal energy E, which presents totally dissipated energy in the system and is a part of total internal energy U.

The state of the system is fixed with fundamental variables $x_1, x_2, \dots, x_n$ that are ascribed to the system, in whatever state (in equilibrium or not) it is. The *non-equilibrium state* of the thermodynamic system is not defined by the fundamental variables unambiguously; one has to add an indefinite number of other variables $\xi_1, \xi_2, \dots$ to describe deviations of the system from the equilibrium state. When the limited list of n fundamental variables is known, the other variables are called *internal variables*.

When the system undergoes external influences, internal variables can emerge and then, being left on their own, they tend to vanish. The main property of the internal variables is their tendency to disappear, which is described by the law (1.16) in linear

doi:10.1088/978-0-7503-3451-8ch2

approximation. The disappearance of internal variables is connected with dissipation of energy of the system, which can be recorded as

$$-\sum_j \Xi_j \, \Delta\xi_j \geqslant 0, \quad \Delta\xi_j \leqslant 0, \tag{2.1}$$

where Ξ_j, $j = 1, 2, \ldots$ are thermodynamic forces that depend, generally speaking, from all variables of the state. The introduction of dissipation of energy allows us to generalize the expression (1.7) for thermal energy of a system for non-equilibrium situations

$$\Delta E = \Delta Q - \sum_j \Xi_j \, \Delta\xi_j + \sum_\alpha \eta_\alpha \, \Delta n_\alpha. \tag{2.2}$$

This equation defines the change of internal thermal energy for the general case, when there are exchanges of substance and energy between the system and environment, that is, for the open system. Recoding equation (2.2), we assume the case of relaxation of internal variables, when $\Delta\xi_j \leqslant 0$, but nothing hinders us from considering the equation also for the case of emergence of internal variables, when increments of internal variables are positive, $\Delta\xi_j \geqslant 0$.

Equation (2.2) describes both increase and decrease in thermal energy of the system due to the changes of internal variables and all other external influences. For the system with invariable composition—closed systems, when there are neither chemical reactions in system, nor streams of substances in or from the system, the last term in (2.2) disappears.

In the situations when the system is isolated and does not fulfill work via fundamental variables, the total internal energy of the systems U does not change, but some internal redistribution of energy occurs, so that potential energy of the system

$$\Delta(U - E) = \sum_i \Xi_i \, \Delta\xi_i \leqslant 0$$

decreases in exactly the same quantity as thermal energy increases. This quantity of thermal energy appears, when the agitated internal variables are relaxing.

For example, the internal potential arises in any case, when one combines two equilibrium systems with different values of temperatures to create a combined system. Whereas total energy of the combined system does not change during the process, values of internal energy of the systems under irreversible transformation are not constant

$$E_1(T_1) + E_2(T_2) \leqslant E_{1+2}(T),$$

where temperature T can be calculated from the condition of conservation of total energy.

2.2 The fundamental principles

2.2.1 Law of conversion and conservation of energy

To define energy of a thermodynamic system U as a function of variables of state, it is necessary to consider the balance of changes of energy under interaction of the system with the environment via the heat received by the system, ΔQ; via the energy received by the system due to variation of a chemical composition (1.2); and via the work of the system through fundamental variables. Collecting all outer influences in the way it was done in section 1.3, one can define the variation of the *internal total energy* of the system

$$\Delta U = \Delta Q - \sum_j X_j \, \Delta x_j + \sum_\alpha (\eta_\alpha + \mu_\alpha) \, \Delta n_\alpha. \tag{2.3}$$

This expression is the same as equation (1.5) and represents the standard form of *the first principle of thermodynamics.*

Equation (2.3) is valid both for reversible and irreversible processes[1]. Under the irreversible processes, external influences bring to existence some internal variables, and the processes of transformation of potential energy connected with internal variables into heat energy occur, which has to be taken into consideration. To take the processes of reallocation of energy into account, we use a parity (2.2), which allows us to record the law of conversion and conservation of energy in the following form

$$\Delta U = \Delta E - \sum_j X_j \, \Delta x_j + \sum_j \Xi_j \, \Delta \xi_j + \sum_\alpha \mu_\alpha \, \Delta n_\alpha. \tag{2.4}$$

The presentation of *the first principle of thermodynamics* in form (2.4) takes into account explicitly the presence of internal processes of redistribution of energy within the system[2]. The internal variables relax and potential energy of internal variables transfers into thermal internal energy of the system. In equilibrium thermodynamics, it is supposed that processes occur so slowly that processes of structurization and destructurization can be neglected.

---

[1] Let us recall that, if the process of transition of a thermodynamic system from one macrostate to another occurs in such a slow way that all characteristic times of processes t are much more than times of relaxation τ_i of internal variables, fundamental variables, in fact, have their equilibrium values, and the process is considered as *reversible*. Otherwise, some times of relaxation appear comparable or greater than characteristic time of process which, in this case, is named *irreversible*. In this case some of the internal variables are emerging and should be taken into account implicitly or explicitly. Further we consider general non-equilibrium states and irreversible processes.

[2] Classical thermodynamics and the theories, which do not use internal variables for the description of non-equilibrium states of the system (Groot and Mazur 1962, Prigogine 1961) include potential energy of internal variables into the total internal energy, so that there is no chance to consider the internal transformations within these theories. However, also in the theories, which use internal variables (Maugin and Muschik 1994), the authors do not separate potential energy of internal variables from internal total energy, which, in our opinion, complicates understating properly the role of internal variables in the description of non-equilibrium processes.

2.2.2 The arrow of time

The introduction of internal variables allows us to explain the mechanism of irreversible evolution of thermodynamic systems—the arrow of time (Pokrovskii 2005, 2013). To do it, we have to considerer processes of emergence and disappearing of internal variables.

Any influences from the environment provoke changes of the state of the system, which, simultaneously with changes of fundamental variables and composition, can bring *emergence of the internal variables*. According to equation (2.4), the quantity

$$\sum_i \Xi_i \, \Delta\xi_i$$

can be considered as an increase in total internal energy due to the increase of internal variables. This part of the total energy, expressed in terms of internal variables, occurs, of course, when the system is affected, as part of the work on the system. Interpreting the work associated with changing of the fundamental and internal variables in equation (2.4) as the work of a thermodynamic system, we can consider the written expression with a minus sign as the work of the system through the internal variables.

Whether there are influences from the environment or not, relaxation of the internal variables always occurs according to the internal laws of the system. The main property of internal variables, described in section 1.4.3, is tending to disappear, while the potential energy of the system is converted to thermal form, and in this situation we have to write

$$\sum_i \Xi_i \, \Delta\xi_i \leqslant 0. \tag{2.5}$$

This relation is valid for the internal variables with any values of relaxation times. However, if the relaxation times of some internal variables are very large (in comparison with characteristic time of the process), these internal variables could be considered as additional fundamental variables. This will be illustrated for the case of a very complex system, similar to living organisms (see chapter 9).

One can see that the emergence of internal variables occurs due to external influences, but disappearance of internal variables is a totally internal process. The processes of relaxation determine impossibility of the reversion of evolution of the thermodynamic system in time and statement (2.5) can be considered as a formulation of *the second principle of thermodynamics*—the formulation, which is, as will be shown in section 2.3.2, equivalent to the formulation of the principle in terms of entropy.

2.2.3 Entropy and temperature

Equation (2.4) defines variations of total internal energy U due to external and internal influences, but they do not define it as a function of state. It is a problem that was already discussed in section 1.3 for the equilibrium situations, and now we are going to describe the non-equilibrium situations. We can believe that entropy keeps its meaning, as a measure of completely degraded energy, both for the equilibrium and non-equilibrium situations, so that we accept that the definition of entropy (1.7) is valid in any case, which allows us to record the definitions of total internal energy and entropy

$$dU = T\,dS - \sum_j X_j\,dx_j + \sum_j \Xi_j\,\Delta\xi_j + \sum_\alpha \mu_\alpha\,\Delta n_\alpha, \tag{2.6}$$

$$dS = \frac{\Delta E}{T}, \quad \Delta E = \Delta Q - \sum_j \Xi_j\,\Delta\xi_j + \sum_\alpha \eta_\alpha\,\Delta n_\alpha. \tag{2.7}$$

Let us stress that the set of internal variables ξ_j in these equations includes all internal variables, both decreasing and increasing.

The introduction of entropy S allows us to define the total internal energy U of the system, as a thermodynamic function of state both for the equilibrium and non-equilibrium situations

$$U = U(S, \boldsymbol{x}, \boldsymbol{\xi}, \boldsymbol{n}).$$

Relations (2.7) define entropy as a characteristic of the state of the system, but, however, does not define this quantity as a function of the system state variables; the introduction of arguments is discussed in section 2.5. The internal energy and some other potential functions will be discussed in section 2.4.

Simultaneously with non-equilibrium entropy S, the temperature T has been introduced by equation (2.7). For equilibrium situations (when values of all internal variables are equal to equilibrium ones), this quantity coincides, apparently, with absolute temperature, and, for non-equilibrium situations, it is a universal characteristic temperature that is absolute temperature for the thermal part of the system with thermal energy E. Of course, the 'real' temperature of the system in a non-equilibrium state could differ from the temperature T that conveniently can be called *reference temperature*. The difference between the 'real' and reference temperatures can be considered as an internal variable and included in the list of $\boldsymbol{\xi}$ (inexplicitly). Also, such a difference can be considered explicitly as internal variables (Muschik 2014).

The concept of temperature still appears fundamental both for equilibrium, and for non-equilibrium situations. By introducing entropy and temperature, as universal characteristics, one provides an adequate description of the thermodynamic system in both equilibrium and non-equilibrium situations. Certainly, in cases of non-equilibrium the situation appears more complex[3], attempts to calculate 'non-

[3] A situation can be rather complicated: temperature can change from point to point in the investigated body (in this case one has to introduce local temperature) or different constituents of the body (partial sets of degrees of freedom) can have different temperatures (for example, the temperature of phonons can be different from the temperature of electrons in a solid), but in any case there is *an empirical characteristic called temperature* which can be ascribed to the entire body, or to its parts (particular subsystems), or to the material point of the body, as a small volume of the system (a point from macroscopic point of view) which, nevertheless, contains a great deal of particles (a thermodynamic system).

equilibrium temperature' according to the rules that are valid for the equilibrium situations, lead to controversial results (Jou and Restuccia 2016).

2.3 Non-equilibrium entropy

The way of extension of the concept of entropy for non-equilibrium states keeps the meaning of entropy, established for the equilibrium situations, as a measure of completely degraded energy in the system, and the equations (2.7) define the change of entropy S for any system as

$$T\,dS = \Delta Q - \sum_i \Xi_i\,\Delta\xi_i + \sum_\alpha \eta_\alpha\,\Delta n_\alpha. \tag{2.8}$$

Thermodynamic force can be defined now as

$$\Xi_j = -T\frac{\partial S}{\partial \xi_j} \tag{2.9}$$

Let us recall, that T is a reference temperature, ΔQ is a flux of heat through borders of the system, and Δn_α represent a total change of substances within the system, whereas $\Delta\xi_i$ is a variation of internal variables.

The fundamental equation (2.8) determines, that various influences on the system lead to variation of the universal quantity—entropy, and generalizes the classical definition of entropy, as a measure of energy in heat form (see section 1.3). Note that different definitions for change of entropy of the non-equilibrium system, which can be found in the literature, if not to be reduced to equation (2.8), do not satisfy the requirement for entropy to be the measure of completely degraded energy.

The definition (2.8) generalizes traditional understanding of entropy. Indeed, if one omits the work of internal variables in this equation, one can, write in accordance with equation (2.5),

$$T\,dS \geqslant \Delta Q + \sum_\alpha \eta_\alpha\,\Delta n_\alpha.$$

The written relation shows that a change of entropy in a real process is more than a change of entropy in a reversible process, and internal production of entropy was invented to compensate the deficiency. This equation in couple with equation (1.6) is the conventional form (see, for example, Maugin and Muschik 1994, equation (3.5)) of equations of traditional thermodynamics without explicit accounting of internal processes.

2.3.1 Entropy as a function of internal variables

The thermodynamic system in a non-equilibrium state is characterized by a set of internal variables ξ_1, ξ_2, ... that, in their entity, describe deviation of the system from the equilibrium state. One can neglect here fluctuations of the thermodynamic quantities and consider the noticeable internal variables, which determine the internal structure of the system. Such internal variables we shall call *complexity internal variables*.

To define entropy as a function of complexity internal variables, let us consider a simple case, when the system does not fulfill work, and external fluxes are absent, so that the change of entropy, according to definition of entropy (2.8), is connected with variations of internal variables and, according to the parities recorded above, can be expressed as

$$dS = -\frac{1}{T}\sum_{j=1}^{s} \Xi_j \, d\xi_j \geqslant 0. \tag{2.10}$$

By virtue of definition (2.9) of the thermodynamic forces, the right part of expression (2.10) represents total differential of entropy as a function of internal variables.

One can consider process of change of entropy of the system from a non-equilibrium state with fixed values of the complexity variables $\xi_1, \xi_2, \ldots \xi_s$ to an equilibrium state, when all complexity variables are equal to zero. If entropy S is considered to be a function of state, the result does not depend on the path of transition, and we can find a difference between values of entropy in equilibrium and non-equilibrium states. The result can be symbolically recorded as

$$S(T, \mathbf{x}, \boldsymbol{\xi}) - S(T, \mathbf{x}, 0) = -\frac{1}{T}\int_0^{\xi_1, \xi_2, \ldots \xi_s} \sum_j \Xi_j \, d\xi_j \leqslant 0. \tag{2.11}$$

Relation (2.11) can be considered as a definition of entropy of non-equilibrium state of a thermodynamic system. One can see from this relation that the value of entropy of a complex system (a system in a non-equilibrium state) is less than value of entropy of the same system in the equilibrium state, assuming that values of temperature and fundamental variables are the same. One can assume that the thermodynamic forces $\Xi_1, \Xi_2, \ldots, \Xi_s$ are known as functions of the complexity internal variables $\xi_1, \xi_2, \ldots, \xi_s$, so that the existence of entropy as a function of internal variables is connected with the possibility to calculate expression (2.11). It is well-known, that only at the certain choice of functions $\Xi_j(\xi_1, \xi_2, \ldots, \xi_s)$, is it possible to calculate the function $S(\xi_1, \xi_2, \ldots, \xi_s)$, however, by virtue of definition of thermodynamic forces (2.9) such an operation is always possible. Examples of calculation of entropy for some simple cases are considered in chapter 4.

2.3.2 Two components of entropy

The total change of entropy of the system dS, defined by expression (2.8), can be split, according to Prigogine (1961), into the two components

$$d\,S = d_e\,S + d_i\,S. \tag{2.12}$$

The first term is a change of entropy influenced by fluxes of heat and substances; the second term is internal production of entropy.

The change of entropy due to fluxes

The entropy changes in the system, due to external influences, $d_e\,S$ is connected with inflow of heat ΔQ and substances through the borders of the system

$$d_e\,S = \frac{1}{T}\left(\Delta Q + \sum_{\alpha} \eta_\alpha\, \Delta^e n_\alpha\right). \tag{2.13}$$

The special index e in the increment of particles $\Delta^e n_\alpha$ shows that the change of particles only due to the inflow through the borders of the system is accounted for. The quantity η_α in equation (2.13) is not chemical potential, but thermal energy coming with the fluxes of substances.

The quantity $d_e\,S$ can be both positive (the flux of heat and/or substances into the system) and negative (the flux of heat and/or substances out of the system). For the case of an *isolated* system, when there are neither heat fluxes, nor matter fluxes between the system and surroundings, $d_e\,S = 0$.

The fluxes of heat and substances into the system or from the system appear at the absence of balance between the system and the environment, which are real fluxes of heat and/or substances through the borders of system. The specific type of the fluxes into the thermodynamic system is determined by a considered problem, but, in any case, it is possible to define empirically a set of streams $J_1, J_2, \dots, J_r$ that are assumed to be given. It is convenient to write the expression for the change of entropy in the system, caused by external fluxes, in the standard form

$$\frac{d_e\,S}{dt} = \frac{1}{T}\sum_{j=1}^{r} K_j J_j, \tag{2.14}$$

where the quantities K_j are characteristics of the system, that is, functions of the temperature, fundamental variables, as well as internal variables. The terms in the sum (2.14) can be both positive, and negative. We shall notice, that the change of entropy of the system due to fluxes of heat and substances differs, generally speaking, from the change of entropy in the environment.

The internal production of entropy

Entropy also changes, due to some internal processes within the system, and this part of the entropy $d_i\,S$ is connected with the change of internal variables

$$d_i S = -\frac{1}{T}\sum_{j} \Xi_j\, \Delta\xi_j, \qquad d_i S \geqslant 0, \tag{2.15}$$

At the process of transition from one macroscopic state to another, some internal variables (that cannot be determined in advance) appear to be agitated and then relax according to internal laws. By virtue of equation (2.5), the quantity $d_i\,S$ can be only non-negative, and this statement is the *standard expression of the second law of thermodynamics*. In reversible processes, when the process occurs in such a slow way (characteristic time of process $t \gg \tau_i$), that all internal variables, in fact, have their equilibrium values, internal production of entropy is equal to zero, $d_i\,S = 0$.

From equation (2.15) the expression for *production of entropy* follows

$$\frac{d_iS}{dt} = -\frac{1}{T}\sum_j \Xi_j \frac{d\xi_j}{dt}, \quad \frac{d_iS}{dt} \geqslant 0. \tag{2.16}$$

The quantity is connected with variation of internal variables. Entropy increases with increase of relaxation of internal variables in the system, so that production of entropy is evidence of the presence of internal variables, which is a sign of internal complexity of the system. Let us recall that internal variables represent every kind of non-equilibrium in the system such as the differences of temperature, concentrations of substances, pressure, etc.

2.4 Potential functions

In the above-described simple situations, entropy can be considered as a potential function for internal variables; there is, however, the necessity to describe the important feature of the thermodynamic system to perform work via fundamental variables in different situations. In this section, the potential functions, as functions of fundamental variables, will be discussed.

2.4.1 Total internal energy

The total internal energy of the system can play a role of some potential function in relation to fundamental variables in situations when entropy of the system is supposed to be a constant (in adiabatic processes). According to the relations (2.6), one can write down the total differential of total internal energy as

$$dU = T\, dS - \sum_j X_j\, dx_j + \sum_\alpha \mu_\alpha\, dn_\alpha + \sum_i \Xi_i\, d\xi_i. \tag{2.17}$$

This expression defines the total internal energy as a function of entropy, composition, fundamental and internal variables

$$U = U(S, \boldsymbol{x}, \boldsymbol{\xi}, \boldsymbol{n}).$$

Thermodynamic forces in the differential (2.17) are determined as partial derivatives of internal energy

$$X_i = -\frac{\partial U}{\partial x_i}, \quad \Xi_i = \frac{\partial U}{\partial \xi_i},$$

The partial derivative of total internal energy over entropy appears to be temperature

$$T = \frac{\partial U}{\partial S}$$

that is *absolute temperature* in the case of equilibrium and *reference temperature* in the case of non-equilibrium.

We recall, that temperature, unlike heat, is a property or parameter of state, which allows us to introduce temperature T as an argument of a certain characteristic function of state.

2.4.2 Free energy

The processes are running often under conditions, when not entropy, but temperature appears to be a controllable variable. To introduce a potential function for such cases (isothermal processes), we address the differential of total energy of the system (2.17) and, following a traditional method, add expression

$$S\,dT - S\,dT$$

to the right-hand side of parity (2.17). Combining the terms, we find the differential of the new potential function $F = U - TS$, named Helmholtz free energy, or simply *free energy*,

$$dF = -S\,dT - \sum_i X_i\,dx_i + \sum_\alpha \mu_\alpha\,dn_\alpha + \sum_i \Xi_i\,\mathrm{d}\xi_i. \tag{2.18}$$

The executed operation—Legandre's transformation—defines *free energy* of the system, as a function of the temperature, composition, fundamental and internal variables

$$F = F(T, \boldsymbol{x}, \boldsymbol{\xi}, \boldsymbol{n}).$$

Thermodynamic forces are defined as partial derivatives of a potential function

$$X_i = -\frac{\partial F}{\partial x_i}, \quad \Xi_i = \frac{\partial F}{\partial \xi_i}, \quad at \quad T = \text{const}.$$

Entropy appears to be the partial derivative of free energy over temperature

$$S = -\frac{\partial F}{\partial T}$$

2.4.3 Enthalpy

To control the heat content within the system, one can introduce the function

$$H = U + \sum_{i=1}^{n} x_i\,X_i,$$

where U is total internal energy, X_j is a force that is affected by the corresponding fundamental variable x_j. The introduced function, according to definition and equation (2.17), has the total differential

$$dH = T\,dS + \sum_i x_i\,dX_i + \sum_\alpha \mu_\alpha\,dn_\alpha + \sum_i \Xi_i\,d\xi_i. \tag{2.19}$$

The first term in this differential represents, according to agreement (see section 2.2.3), a change of internal thermal energy within the open system in the non-equilibrium situation, so that the function H presents content of heat in the system even for the case, when variables X, ξ and n are constant.

Let us recall that in the simplest case the middle term in the right side of the equation is $V\ dp$ (V is a volume of the system, p is pressure inside the volume). In this case, the function H can be used to describe processes under controlled entropy and pressure.

2.4.4 Gibbs free energy

For description of processes, which are running at given temperature and 'forces' (in the simplest case, pressure), a function called Gibbs free energy or thermodynamic potential (Landau and Lifshitz (1986) can be introduced)

$$G = H - TS = F + \sum_i x_i\, X_i = U - TS + \sum_i x_i\, X_i.$$

The function, according to the definition and previous expressions for differentials of functions U, F and H has total differential

$$dG = -S\ dT + \sum_i x_i\ dX_i + \sum_\alpha \mu_\alpha\ dn_\alpha + \sum_i \Xi_i\ d\xi_i. \tag{2.20}$$

In the simplest case, when the second term in the right side of the equation is reduced to $V\ dp$ (V is a volume of the system, p is pressure inside the volume), the function G depends on temperature T and pressure p (apart from the variables ξ and n), and this thermodynamic potential is used for description of the chemical reaction that is running under controlled temperature and pressure (see chapter 6).

2.5 Entropy as a function of state

In virtue of relations (2.18) and (2.20), entropy can be defined as a partial derivative of free energy or Gibbs free energy over temperature

$$S = -\frac{\partial F}{\partial T} = -\frac{\partial G}{\partial T}. \tag{2.21}$$

These definitions allow us to introduce entropy as a function of state variables. The first of relations (2.21), for example, defines entropy as a function of composition, temperature, fundamental and internal variables

$$S = S(T, \boldsymbol{x}, \boldsymbol{\xi}, \boldsymbol{n}).$$

The total differential of entropy for this case can be compared with the change of entropy of the system (2.8), which defines some of the partial derivatives of entropy, so that the total differential can be written as

$$\mathrm{d}S = \frac{\partial S}{\partial T}\mathrm{d}T + \frac{\partial S}{\partial \boldsymbol{x}}\mathrm{d}\boldsymbol{x} - \frac{1}{T}\boldsymbol{\Xi}\ \mathrm{d}\boldsymbol{\xi} + \frac{1}{T}\eta\ \mathrm{d}\boldsymbol{n}.$$

The partial derivatives of entropy with respect to temperature T and fundamental variables $\boldsymbol{x}$ are connected with increment of heat by relation

$$\frac{1}{T}\Delta Q = \frac{\partial S}{\partial T}\mathrm{d}T + \frac{\partial S}{\partial \boldsymbol{x}}\mathrm{d}\boldsymbol{x}.$$

The above equations allow us to define the total derivative of entropy, according to equations (2.8), (2.14) and (2.16), in the form

$$\frac{\mathrm{d}S}{\mathrm{d}t} = -\frac{1}{T}\sum_{j=1}^{s}\Xi_j\frac{\mathrm{d}\xi_j}{\mathrm{d}t} + \frac{1}{T}\sum_{j=1}^{r}K_j J_j. \tag{2.22}$$

The last term in equation (2.22) contains heat fluxes and fluxes of substances through the borders of the system; changes of substances within the system are controlled by internal variables (see the situation for the systems with chemical reactions in chapter 6). The quantities Ξ and K_j in equation (2.22) are characteristics of the system, that is, functions of the composition, temperature, fundamental variables, as well as internal variables. The incoming flux of entropy is different, generally speaking, from the flux of entropy, coming out of the environment.

In the situations close to equilibrium, values of internal variables are small, so that entropy can be presented as expansion into the series

$$S = S_0 - \frac{1}{2}\sum_{i,j=1}^{\infty}S_{ij}(0)\,\xi_i\xi_j + \cdots, \qquad S_{ij}(0) = \left(\frac{\partial S}{\partial\xi_i\partial\xi_j}\right)_{T,\boldsymbol{x},\xi=0}. \tag{2.23}$$

The expansion does not contain terms of the first order, so the thermodynamic forces Ξ_j disappear in equilibrium situations. In virtue of equation (2.16), value of entropy of a system in a non-equilibrium state is less than value of entropy of the same system in the equilibrium state, so that the matrix S must be non-negatively determined. Components of the matrix S are functions of temperature and fundamental variables.

Equation (2.23) allow us to record expression for thermodynamic force near the equilibrium state. It is possible to see, that, in linear approximation, thermodynamic forces are connected linearly with internal variables

$$\Xi_j = -T\frac{\partial S}{\partial\xi_j} = T\,S_{jk}\xi_k + \cdots \tag{2.24}$$

Let us recall, that we assume the values of internal variables to be zero at the equilibrium state, so that the thermodynamic forces Ξ_j disappear in equilibrium situations.

2.6 Chemical potential

Considering the thermodynamic system as a certain amount of substances, we could be interested in how the thermodynamic characteristics of the system change with the changes of the number of particles (moles). One can notice that, according to equations (2.17)–(2.20) the numbers of particles (moles) n_j can be considered as arguments of the thermodynamic characteristics, which is going up to Gibbs (1875) and his followers (see, for example, Landau and Lifshitz 1986, section 24). The quantities that describe changes of the potential functions on the numbers n_j are defined as total derivatives of thermodynamic functions over the number of particles

$$\mu_\alpha^{(U)} = \frac{\mathrm{d}U}{\mathrm{d}n_\alpha}, \quad \mu_\alpha^{(F)} = \frac{\mathrm{d}F}{\mathrm{d}n_\alpha}, \quad \mu_\alpha^{(H)} = \frac{\mathrm{d}H}{\mathrm{d}n_\alpha}, \quad \mu_\alpha^{(G)} = \frac{\mathrm{d}G}{\mathrm{d}n_\alpha}. \tag{2.25}$$

Considering that the all fundamental and internal variables do not depend on the constitution, definitions (2.25) in couple with equations (2.17)–(2.20) define two characteristic quantities: chemical potential and specific enthalpy, correspondingly,

$$\mu_\alpha = \mu_\alpha^{(F)} = \mu_\alpha^{(G)}, \quad h_\alpha = \mu_\alpha^{(U)} = \mu_\alpha^{(H)} = \mu_\alpha + T\frac{\mathrm{d}S}{\mathrm{d}n_\alpha}. \tag{2.26}$$

By virtue of these definitions, the quantity η_α, due to equation (2.8), can be interpreted as a contribution of unity of substance into thermal energy of the system $\eta_\alpha = T\frac{\mathrm{d}S}{\mathrm{d}n_\alpha}$; the quantity, that was initially introduced by equation (1.7), can be defined again as

$$\eta_\alpha = h_\alpha - \mu_\alpha. \tag{2.27}$$

2.7 The peculiarities of thermodynamic functions

From the point of view of mechanics, macroscopic features of the behavior of a system of many particles are mysterious (Prigogine 1961). In any case, the behavior of a thermodynamic system is not reduced to any principles of mechanics. Indeed, in a composite isolated system, as described in sections 1.4.1 and 1.4.2, there is an alignment in concentration and temperature under other characteristics to be constant, which means that the total energy of the system does not fully control the observed macrodynamics. The characteristic that controls the described phenomena is entropy. The tendency of a thermodynamic system to an equilibrium state is described as the tendency of entropy to a maximum value, and is ultimately related to the behavior of internal variables that, left to themselves, tend to disappear. Here we are not talking about fluctuations, but about noticeable internal variables, which are called *internal complexity variables*.

Analysis of behaviour of thermodynamic systems allows us to define main thermodynamic characteristics of the system, among them the total internal energy, free energy and entropy, correspondingly,

$$\begin{aligned} U(S, \boldsymbol{x}, \boldsymbol{\xi}, \boldsymbol{n}) &= U(S, x_1, x_2, \dots, x_n; \xi_1, \xi_2, \dots, \xi_s; n_1, n_2, \dots, n_k), \\ F(T, \boldsymbol{x}, \boldsymbol{\xi}, \boldsymbol{n}) &= F(T, x_1, x_2, \dots, x_n; \xi_1, \xi_2, \dots, \xi_s; n_1, n_2, \dots, n_k), \\ S(T, \boldsymbol{x}, \boldsymbol{\xi}, \boldsymbol{n}) &= S(T, x_1, x_2, \dots, x_n; \xi_1, \xi_2, \dots, \xi_s; n_1, n_2, \dots, n_k). \end{aligned} \tag{2.28}$$

By virtue of the way of introduction, these functions and temperature, irrespective of the presence or absence of internal variables, are connected by universal relations

$$F = U - TS, \quad S = -\frac{\partial F}{\partial T}, \quad T = \frac{\partial U}{\partial S}. \tag{2.29}$$

By virtue of equation (2.16), value of entropy of the systems has the maximum value in a state of equilibrium. From this assertion and definitions of characteristic functions of the system, it follows that in this point of equilibrium, internal thermal energy gets the greatest possible value, and total and free energy are minimal in relation to internal variables. These properties provide stability of state of equilibrium. The tendency of entropy to a greater value was declared as a special principle—the second law of thermodynamics. This principle, if we postulate the existence of entropy, as a function of state with certain properties, allows us to describe the phenomena, for example, of the equality of temperatures and chemical potentials in equilibrium (see, for example, Landau and Lifshitz 1986, pp 50–52).

In situations close to equilibrium, values of internal variables are small, so that thermodynamic functions can be presented in the form of expansion to series. For entropy, decomposition in the elementary approximation is presented by expression (2.23).

References

Gibbs J W 1875 On the equilibrium of heterogeneous substances *Trans. Conn. Acad. Arts Sci.* **III** 108–248 343–524

Gibbs J W 1928 *Equilibrium of non-homogenous substances, 1895–1898: Collected works* (New York, London: Longman Green)

Groot de S R and Mazur P 1962 *Non-Equilibrium Thermodynamics* (Amsterdam: North-Holland)

Jou D and Restuccia L 2016 Caloric and entropic temperatures in non-equilibrium steady states *Physica* A **460** 246–53

Landau L D and Lifshitz E M 1986 *Statistical Physics* vol 1 3rd edn (Oxford: Pergamon)

Maugin G A and Muschik W 1994 Thermodynamics with internal variables part I. General concepts *J. Non-Equil. Thermodyn.* **19** 217–49

Muschik W 2014 Contact temperature and internal variables: a glance back, 20 years later *J. Non-Equil. Thermodyn.* **39** 113–21

Pokrovskii V N 2005 Extended thermodynamics in a discrete-system approach *Eur. J. Phys.* **26** 769–81

Pokrovskii V N 2013 A Derivation of the main relations of non-equilibrium thermodynamics *ISRN Thermodyn.* **2013** 906136

Prigogine I 1961 *Introduction to Thermodynamics of Irreversible Processes* 2nd edn (New York: Interscience)

Chapter 3

Dynamics of internal variables

In the case of reversible processes, changes of fundamental variables follow external influences and all internal variables can be neglected. In a general case, to describe the actual situation, one has to consider emerging and evolution of the internal variables, which are not influenced by the external forces directly. As an example, we can point to a situation, when temperature of the environment increases and heat starts to get into the thermodynamic system. In this situation, gradients of temperature, that is, in our interpretation, internal variables, are emerging, and internal processes appear: the system tends to a new equilibrium, but the state of the system is determined by a play between external influences and internal processes. To move further, our speculations should be completed by equations for evolution of internal variables.

3.1 Internal variables of complexity

Let us consider a thermodynamic system in a non-equilibrium state that is completely determined by temperature T, fundamental and internal variables, $\boldsymbol{x}$ and $\boldsymbol{\xi}$. Among the internal variables, there are some internal variables that are noticeable and play a distinctive role in the macroscopic description of the system as, for example, gradients of temperature or deviations of concentration of different substances from equilibrium values. Such variables can be called *complexity variables*, their number is restricted $\xi_1, \xi_2, \ldots, \xi_s$. Except for variables of complexity, there can be an uncertain number of some other internal variables $\xi_{s+1}, \xi_{s+2}, \ldots$, which, in contrast to the complexity internal variables, cannot be controlled. The

doi:10.1088/978-0-7503-3451-8ch3

fluctuations of thermodynamic quantities are considered as uncontrollable internal variables (Groot and Mazur 1962, Landau and Lifshitz 1986, chapter 12)[1].

The real systems are being found usually in non-equilibrium states, and, for the description of the non-equilibrium, one needs the introduction of a set of internal complexity variables ξ_1, ξ_2, ... To support the non-equilibrium state of a complex system, fluxes of substances and heat energy are needed to flow through borders of the system, so that the complex system is an open system with a set of fluxes of, generally speaking, energy and substances J_1, J_2, ... , J_r. The fluxes create and support a non-equilibrium state of the system that returns eventually to equilibrium, if the external forces disappear.

The internal variable, as any thermodynamic quantity, appears to be fluctuating variable, and in order to be a complexity variable, a quantity must have fluctuations less than the quantity itself. Each complexity internal variable can be presented as a sum of the regular (averaged in some way) and random components, and the equation for complexity internal variables should include random forces. Further on, only regular components of variables are considered. Discussion of random components, also as well as fluctuations of other thermodynamic variables, which can be considered as internal variables as well, is omitted. For expansion of the description, it is possible to use the mathematical apparatus of stochastic non-linear non-equilibrium thermodynamics described, for example, by Stratonovich (1994).

3.2 Preferable values of internal variables

A change of entropy of the system $S(T, \boldsymbol{x}, \boldsymbol{\xi}, \boldsymbol{n})$ occurs due to changing of arguments of the function that are influenced by the incoming and outcoming flows of heat and substances, which is the contribution defined by expression (2.8). We assume (Pokrovskii 2013) that the tendency of entropy to acquire the greatest value is kept, and, similar to the case of isolated systems, when the maximum value of entropy determines the equilibrium state, extreme values of entropy of the system in non-equilibrium under the presence of external influences defines special points in state space. These points are defined by a relation obtained by equating the total

[1] The probability of deviation of internal variables from zero values can be written as

$$w(\xi_1, \xi_2, \ldots) = C \exp[\Delta S(\xi_1, \xi_2, \ldots)],$$

where C is a normalization constant. The quantity ΔS is the difference between the entropy values in the non-equilibrium and equilibrium states, which is defined by equation (2.23), so that the probability of small deviation of fluctuations from the zero value is determined by the simple formula

$$w(\xi_1, \xi_2, \ldots) = C \exp\left(-\frac{1}{2}\sum_{i,j} S_{ij}\xi_i\xi_j\right).$$

We do not include here fluctuations in the thermodynamic system (local deviations of density, temperature, concentration of substances, and other values from the equilibrium zero values), which can be interpreted as a huge number of non-zero uncontrolled internal variables.

derivative of entropy to zero. We use the definition of the total derivative (2.22) to write the equation for the preferable values

$$-\frac{1}{T}\sum_{j=1}^{s}\Xi_j\frac{\mathrm{d}\xi_j}{\mathrm{d}t}+\frac{1}{T}\sum_{j=1}^{r}K_j\,J_j=0. \tag{3.1}$$

The quantities Ξ_j and K_j in equation (3.1) are characteristics of the system, that is, functions of the composition, temperature, fundamental variables, as well as internal variables. The incoming flux of entropy is different, generally speaking, from the flux of entropy, coming out of the environment.

The relation (3.1) at given temperature, fundamental variables and external fluxes can be considered as an equation for unknown internal variables, which, in this case, determines *preferable values of internal variables* $\boldsymbol{\xi}^\circ$. The state variables T, $\boldsymbol{x}$, $\boldsymbol{\xi}$, $\boldsymbol{n}$ change under external influences, that is, under fluxes of heat and substances from the environment and/or with alternation of fundamental variables. The rate of change of internal variables in a point of quasi-equilibrium can be replaced by rate of change of preferable variables (see equation (3.4))

$$\left.\frac{\mathrm{d}\boldsymbol{\xi}}{\mathrm{d}t}\right|_{\boldsymbol{\xi}=\boldsymbol{\xi}^\circ}=\frac{\mathrm{d}\boldsymbol{\xi}^\circ}{\mathrm{d}t}.$$

This allows, using also the definition of thermodynamic force, to write the equation for preferable values as

$$\sum_{j=1}^{s}\Xi_j(\boldsymbol{\xi}^\circ)\frac{\mathrm{d}\xi_j^\circ}{\mathrm{d}t}=\sum_{j=1}^{r}K_j(\boldsymbol{\xi}^\circ)J_j. \tag{3.2}$$

The preferable values of internal variables, corresponding to the actual values of other variables and external influence, are determined, apparently, by history of the application of external influences. Note that the preferable values of variables describing fluctuations are always considered to be equal to zero, so that the number of preferable variables in equation (3.2) is restricted.

In the general case, equation (3.2) cannot unambiguously define preferable values. However, the situation is not so hopeless, when considering special cases. If, for example, temperature and fundamental variables of the system do not change and external influences are given by constant fluxes of heat and substances into the system J_1, J_2, … , J_s, it is possible to make some assumptions. We consider, that each flux corresponds to only one of the internal variables, so that the number of distinctive preferable variables is equal to the number of fluxes, that is, $\xi_l^\circ \neq 0$ $l = 1, 2, \ldots, s$ and $\xi_l^\circ = 0$ $l = s+1, s+2, \ldots$. Taking the arbitrariness in the fluxes also into account, equation (3.2) can be reduced to the set of equations

$$\frac{d\xi_i^\circ}{dt}=B_iJ_i,\quad i=1,2,\ldots,s. \tag{3.3}$$

The coefficients B_i are functions of variables of state, but, in the simplest case, they have constant values that can be settled by a choice of fluxes in respect to a choice of internal variables. When an 'appropriate' choice of fluxes in the form of $J_j \sim \xi_j/\tau_j$, where τ_j is a time of relaxation, has been made, the quantities B_i have numerical values. Then, by a choice of the internal variables and the fluxes, it is possible to reduce values B_i to units, whose approximation will be used further.

3.3 Evolution of internal variables

The actual values of the internal variables ξ_1, ξ_2, … differ from the preferable values. The deviations of internal variables from their current preferable values $\xi_i - \xi_i^\circ$, ($i = 1, 2, \ldots$) determine the tending of the internal variables to preferable values. The change of internal variables ξ_1, ξ_2, … is determined by internal laws of movement of particles of the thermodynamic system, so that the equation for a change of the internal variables can be written in the general form

$$\frac{d(\xi_i - \xi_i^\circ)}{dt} = -R_{ij}(T, x_1, x_2, \ldots, x_n; \xi_1, \xi_2, \ldots; \xi_1^\circ, \xi_2^\circ, \ldots)\left(\xi_j - \xi_j^\circ\right), \quad i = 1, 2, \ldots. \tag{3.4}$$

This equation describes evolution of internal variables to preferable values $\boldsymbol{\xi}^\circ$, which are defined by equation (3.2), in general form, or, for simple situations, by equation (3.3). The right-hand side of equation (3.4) describes a change of the variables, according to internal laws of the system, the external influences are presented via the preferable values[2]. The written equation describes that preferred values ξ_i° are points of attraction. The 'minus' sign is chosen for convenience: matrix R_{ij} in the situations close to equilibrium (at $\xi_i^\circ = 0,\ i = 1, 2, \ldots$) is positive definite in many simple cases, though there are some important exclusions from the requirements for matrix R_{ij} to be positive definite. As for example, one can look at the thermodynamic systems with oscillating chemical reactions in section 6.3.1. One can assume that the matrix R_{ij} allows one to choose such a combination (transformation) of variables, that the dynamic equation (3.4) for these variables transforms into a set of independent equations for each variable. Regarding this case, the right-hand side of equation (3.4) can be expanded in series over powers, so that the evolution equation can be reduced to a relaxation equation for each internal variable

$$\frac{d(\xi_i - \xi_i^\circ)}{dt} = -\frac{1}{\tau_i}(\xi_i - \xi_i^\circ), \quad i = 1, 2, \ldots, \tag{3.5}$$

where $\tau_i = \tau_i(T, x_1, x_2, \ldots, x_n) > 0$ is a relaxation time of corresponding variables. Changing according to the internal laws of motion, the excited internal variables tend to their unique equilibrium value, so that the system tends to its equilibrium state.

[2] Let us note that the internal variables are depending generally on the space coordinates $\xi_i = \xi_i(t, x, y, z)$, so that the equations for change of internal variables can contain also terms (omitted here) for the description of processes of diffusion of internal variables. An example of such a kind is described in section 6.3.

References

Groot de S R and Mazur P 1962 *Non-Equilibrium Thermodynamics* (Amsterdam: North-Holland)
Landau L D and Lifshitz E M 1986 *Statistical Physics* vol 1 3rd edn (Oxford: Pergamon)
Pokrovskii V N 2013 A derivation of the main relations of non-equilibrium thermodynamics *ISRN Thermodyn* **2013** 906136
Stratonovich R L 1994 *Nonlinear Nonequilibrium Thermodynamics* (Heidelberg: Springer)

Chapter 4

Non-equilibrium steady states

All thermodynamic functions, and also complexity internal variables of the system, being in a stationary state, are constant, though internal motion within the system does not stop. The dynamic equilibrium of the system in stationary non-equilibrium state is supported with fluxes of heat and substances. Though the set of internal variables are specific for each physical situation, the general consideration of such states can be presented, as it is also possible to do for the equilibrium state. In this chapter, the simple case, when the non-equilibrium state of the system is close to equilibrium, will be considered and the results of so-called *linear irreversible thermodynamics* will be reproduced.

4.1 Dynamics of internal variables

Apart from temperature T, the state of any thermodynamic system is specified with a set of fundamental variables $x_1, x_2, \dots, x_n$, a set of distinctive complexity internal variables $\xi_1, \xi_2, \dots, \xi_s$ and an uncertain number of some other internal uncontrollable variables $\xi_{s+1}, \xi_{s+2}, \dots$.

In a steady state, according to the definition, all thermodynamic functions are constant, complexity internal variables are also constant, however, we must not hurry to equate the derivative of the thermodynamic characteristics to zero. The issue is that in any stationary state there is internal motion. As for the complexity internal variables, there is the balance of processes of excitation (due to external forces) and relaxation of internal variables, and these processes do not stop in the steady state. Under the simple linear approximation, the process of excitation can be described by equation (3.3), that is,

$$\frac{d\xi_i^\circ}{dt} = B_i J_i, \quad i = 1, 2, \dots, s. \tag{4.1}$$

doi:10.1088/978-0-7503-3451-8ch4

The quantities B_i are considered as units in the following. Evolution of internal variables near the selected stationary point can be written on the base of equations (3.4) as

$$\frac{d\xi_i}{dt} = J_i - R_{il}(\xi_l - \xi_l^\circ) - \frac{\partial R_{ik}}{\partial \xi_l}(\xi_k - \xi_k^\circ)(\xi_l - \xi_l^\circ) + \cdots, \qquad (4.2)$$
$$i = 1, 2, \dots, s.$$

The positive definite matrix R_{il} and its derivatives are defined in the considered preferable steady-state point $\xi_1^\circ, \xi_2^\circ, \dots, \xi_s^\circ$.

In a steady state, the values of internal variables are constant, but this is achieved under the balance of two processes: relaxation and excitation of internal variables by external flows. Considering the processes of relaxation and excitation in a stationary situation to be independent, in correspondence with equation (4.2), one can record two parities in the elementary approximation

$$\frac{d\xi_i}{dt} = -R_{ij}\, \xi_j, \quad J_i = -R_{ij}\, \xi_j^\circ, \quad i = 1, 2, \dots, s. \qquad (4.3)$$

The matrix R_{jk} depends on stationary values of internal variables ξ_j° ($j = 1, 2, \dots$). We shall note, that the second set of equations in (4.3) can be specified as empirical parities, examples of which are parities (1.9) and (1.12), which are valid, certainly, only for the steady-state case in linear approximation.

On some reasonable assumptions about properties of the matrix R_{jk}, from the second set of equations (4.3) it follows that systems of the values of internal variables in steady states are determined by temperature, fundamental variable and external fluxes

$$\xi_j^{(k)} = \xi_j^{(k)}(T, x_1, x_2, \dots, x_n; J_1, J_2, \dots, J_r),$$
$$j = 1, 2, \dots, s, \quad k = 1, 2, \dots, m.$$

Assuming, that a few solutions of equations (4.3) are possible, we have supplied the solutions with the index of steady state $k = 1, \dots, m$. The number of steady states m depends on number of fluxes r. At $r > 0$, there can be some stationary points which are distinct from the equilibrium point. Number 0 is kept for a designation of *the equilibrium macrostate* of the systems.

4.2 Entropy near a steady state

To estimate behaviour of a thermodynamic system near a stationary point, we shall specify dependence of entropy on internal variables. We separate, in this case, the internal variables locating steady state of a system (complexity variables). An expansion of entropy of the thermodynamic system near a preferable (steady-state) point begins with linear terms

$$S(\boldsymbol{\xi}) = S(\boldsymbol{\xi}^\circ) - \sum_{j=1}^{s} S_j(\xi_j - \xi^\circ{}_j) - \frac{1}{2}\sum_{i,j=1}^{\infty} S_{ij}\,(\xi_i - \xi^\circ{}_i)(\xi_j - \xi^\circ{}_j) + \cdots$$

$$S_j = -\left(\frac{\partial S}{\partial \xi_j}\right)_{T,x,\boldsymbol{\xi}^\circ}, \quad j = 1, 2, \ldots, s, \tag{4.4}$$

$$S_{ij} = -\left(\frac{\partial^2 S}{\partial \xi_i \partial \xi_j}\right)_{T,x,\boldsymbol{\xi}^\circ} \quad i, j = 1, 2, \ldots$$

It is taken into account, that, apart from the complexity internal variables $\xi_1, \xi_2, \ldots, \xi_s$, there is a set of internal variables with numbers $s+1, s+2, \ldots$, that describe deviations of the system from the stationary state, but in contrast to complexity internal variables they are uncontrolled and present fluctuations. Matrices S_j and S_{lk} are calculated not in equilibrium point, as in expression (2.23), but in the stationary point and depend, apart from temperature and fundamental variables, on the preferable values of complexity internal variables.

As a function of the internal complexity variables, entropy has no features that would allow us to characterize the stationary state. Considering the variables of complexity to be fixed, expansion of entropy, as a function of all other possible internal variables, is reduced to the form

$$S(\boldsymbol{\xi}) = S(\boldsymbol{\xi}^\circ) - \frac{1}{2}\sum_{i,j=s+1}^{\infty} S_{ij}\,\xi_i\xi_j + \cdots,$$

$$S_{ij} = -\left(\frac{\partial^2 S}{\partial \xi_i \partial \xi_j}\right)_{T,x,\boldsymbol{\xi}^\circ} \quad i, j = s+1, s+2, \ldots \tag{4.5}$$

It is assumed that a stationary state of thermodynamic system near the equilibrium state is stable, so that matrix $S_{lk}(\xi_1^\circ, \xi_2^\circ, \ldots, \xi_s^\circ)$ in expression (4.5) is positive definite: entropy, as a function of internal variables, in the fixed stationary state has a maximum. The properties of matrix S_{lk} in a stationary point, which is far from equilibrium, remain uncertain.

Relation (4.4) allows us to write an expression for thermodynamic forces in a point near the stationary state

$$\Xi_j(\boldsymbol{\xi}) = -T\frac{\partial S}{\partial \xi_j} = \Xi_j(\boldsymbol{\xi}^\circ) + T\,S_{jl}(\xi_l - \xi^\circ{}_l) + \cdots, \quad j = 1, 2, \ldots, s. \tag{4.6}$$

The first terms of the expansion of the thermodynamic forces are constant. In the situations close to the equilibrium state, in the simplest approach, thermodynamic forces are connected linearly with internal variables

$$\Xi_j = -T\frac{\partial S}{\partial \xi_j} = T\,S_{jk}\xi_k + \cdots \tag{4.7}$$

Let us recall, that, in the state of equilibrium, values of internal variables are considered to be zero, and thermodynamic forces Ξ_j disappear.

4.3 Production of entropy

In a stationary non-equilibrium state, values of thermodynamic functions of a system, including entropy, are constant. The production of entropy inside the system (2.16) and corresponding decrease in entropy of the system (2.14), due to fluxes of heat and/or substances, are balanced

$$\frac{d_i S}{dt} = -\frac{d_e S}{dt}. \tag{4.8}$$

With help of the definition of production of entropy (2.16), equations (4.2) and (4.7) allow us to write an expansion of the quantity near a steady-state point

$$\begin{aligned} \frac{d_i S}{dt} = & -\frac{1}{T}\sum_{j=1}^{s} \Xi_i J_i + \frac{1}{T}\sum_{j,l=1}^{s}\left(-TS_{jl}J_j + \Xi_j R_{jl}\right)(\xi_l - \xi_l^{\circ}) \\ & + \sum_{j,k,l=1}^{s}\left(S_{jk}R_{jl} + \Xi_j \frac{\partial R_{jk}}{\partial \xi_l}\right)(\xi_k - \xi^{\circ}{}_k)(\xi_l - \xi^{\circ}{}_l) \\ & + \sum_{j,k,l=s+1}^{\infty} S_{jk}R_{jl}\xi_k\, \xi_l + \cdots \end{aligned} \tag{4.9}$$

The values of all matrices are determined in the considered steady-state point $\xi_1^{\circ}, \xi_2^{\circ}, \ldots, \xi_s^{\circ}$.

In the simplest approximation, the expression for production of entropy can be written as

$$\frac{d_i S}{dt} = -\frac{1}{T}\sum_{j=1}^{s} \Xi_j J_j. \tag{4.10}$$

This expression represents production of entropy in the conventional form as the sum of products of fluxes and thermodynamic forces. The 'minus' sign in expression (4.10) is usually omitted, although it is necessary, since the signs of fluxes are always opposite to the signs of forces (internal variables).

Equation (4.10) is considered as one of the basic statements of non-equilibrium thermodynamics (Prigogine 1961, Groot and Mazur 1962, Maugin and Muschik 1994, Kondepudi and Prigogine 1999, Jou *et al* 2001). One can note, that representation (4.10) is valid only for steady-state situations and for small deviations from equilibrium state.

Comparison of equation (2.14) and (4.10), when relation (4.8) is valid, determines that quantity K_j in equation (2.14) coincides with thermodynamic force, $K_i = \Xi_i, \quad i = 1, 2, \ldots, s$ for stationary states.

4.4 On the criterion of stability of stationary states

Expression (4.9) shows, that, in a steady-state point, production of entropy, as a function of the internal variables of complexity, has no peculiar points. However, if the steady-state point is fixed ($\xi_l = \xi^\circ_l,\ l = 1, 2, \ldots s$), expression (4.9) takes the form

$$\frac{d_i S}{dt} = -\frac{1}{T}\sum_{j=1}^{s} \Xi_i J_i + \sum_{j,k,l=s+1}^{\infty} S_{jk} R_{jl} \xi_k \xi_l + \cdots$$

The matrices R_{jk} and S_{lk} depend on the stationary values of the internal variables ξ°_j ($j = 1, 2, \ldots$). The behaviour of production of entropy in vicinity of a steady-state point is determined by terms of the second order with respect to all internal variables, excepting the complexity variables. The coefficients are comprised by the product of two matrices R_{jk} and S_{lk} that were originated from equations (3.4) and (4.4) as positive-definite matrices in states close to equilibrium. The matrices are calculated in the steady-state point and depend, apart from temperature and fundamental variables, on the complexity internal variables.

If the combination $S_{lk}R_{ij}$ is positive-definite, the steady-state point is stable, that is, having been allowed to be on their own, the perturbed internal variables are tending to some fixed values, so that the system is tending to a nearest steady-state point. At the equilibrium point ($\xi^\circ_l = 0,\ l = 1, 2, \ldots s$) entropy production has a minimum, since the second-order term contains the product of two positive-definite matrices and is a positive-definite quadratic form, so that entropy production has a minimum. Due to the continuity of the components of the matrices R_{jk} and S_{lk} as functions of fixed internal variables, the situation persists for stationary points near the equilibrium state, which confirms the validity of Prigogine's minimum entropy production principle for the equilibrium point and its surroundings (Prigogine 1961, Kondepudi and Prigogine 1999). At points that are far from the equilibrium point matrices R_{jk} and S_{lk} are not necessarily positive definite. Moreover, it is stated that the stability of stationary states at points far from equilibrium is characterized by a maximum of entropy production (Hackl and Fischer 2008, Kleidon 2009).

Entropy production can be considered as a characteristic function for stationary states, and stability is associated with the presence of an extremum of entropy production.

4.5 Relationship between fluxes and thermodynamic forces

The expressions for production of entropy (2.16) and (4.10) can be reformulated with the help of equations (4.3) and (4.7) to get the relation

$$\frac{1}{T}\sum_{j,k=1}^{s} \Xi_j R_{jk}\, \xi_k = -\sum_{j,k=1}^{s} S_{jk}\xi_k\, J_j. \tag{4.11}$$

The written equation, by virtue of the supposed arbitrariness and independence of internal variables, is followed by a relation between fluxes and thermodynamic forces

$$\frac{1}{T}\sum_{j=1}^{s} \Xi_j R_{jk} = -\sum_{j=1}^{s} S_{jk}\, J_j. \tag{4.12}$$

This relation can be rewritten as

$$J_i = -L_{ik}\,\Xi_k, \qquad L_{ik} = \frac{1}{T}R_{il}S_{lk}^{-1}. \tag{4.13}$$

Considering the linear approach, the components of matrices R_{jk} and S_{lk} are constants, but in the more general case it is necessary to consider them as functions of internal variables. By virtue of definition, the matrix L is positive definite.

The relations between fluxes and thermodynamic forces in linear approximation are fundamental relations of the non-equilibrium thermodynamics (Prigogine 1961, Groot and Mazur 1962, Maugin and Muschik 1994, Kondepudi and Prigogine 1999, Jou *et al* 2001). We have shown, that these parities are a consequence of the main principles of thermodynamics and valid under two conditions: first, deviations from the equilibrium state are small, and, second, the state is stationary.

Let us note in addition, that there is a statement (Prigogine 1961, Groot and Mazur 1962, Landau and Lifshitz 1986, Maugin and Muschik 1994, Kondepudi and Prigogine 1999, Jou *et al* 2001) that matrix L is symmetric or antisymmetric

$$L_{ij} = \pm L_{ji} \tag{4.14}$$

For the proof of this statement, usually referred to as *Onsager principle*, one has to address other principles and some assumptions considered in the following section.

4.5.1 Symmetry of kinetic coefficients

The proof of the symmetry of kinetic coefficients is based on the property of invariance of correlations of fluctuations of various quantities with respect to reversion of time, which is fair for equilibrium situations (Prigogine 1961, Groot and Mazur 1962, Landau and Lifshitz 1986). Being interested in stationary states, it is possible to consider fluctuations of internal variables near their stationary values and to assume, that time correlations of random deviations of internal variables also are invariant with respect to reversion of time. In other words, for correlations of various quantities, it is possible to write a relation

$$\langle(\xi_i - \xi_i^\circ)_t(\xi_k - \xi_k^\circ)_0\rangle = \pm\langle(\xi_k - \xi_k^\circ)_t(\xi_i - \xi_i^\circ)_0\rangle. \tag{4.15}$$

The 'minus' sign arises in the case, when the internal variable itself changes the sign at the reversion of time.

Further, one can take advantage of the equation of evolution (3.4), which, with the use of expression for thermodynamic force (4.6) and definition of a matrix of kinetic coefficients (4.13), can be presented in the form

$$\frac{d(\xi_i - \xi_i^\circ)}{dt} = -L_{il}\,(\Xi_l(\boldsymbol{\xi}) - \Xi_l(\boldsymbol{\xi}^\circ)). \tag{4.16}$$

This equation allows one, after differentiation of relation (4.15), to write down

$$L_{il}\,\langle(\Xi_l(\boldsymbol{\xi}) - \Xi_l(\boldsymbol{\xi}^\circ))(\xi_k - \xi_k^\circ)_0\rangle = \pm L_{kl}\,\langle(\Xi_l(\boldsymbol{\xi}) - \Xi_l(\boldsymbol{\xi}^\circ))(\xi_i - \xi_i^\circ)_0\rangle. \tag{4.17}$$

By virtue of supposed independence of deviations of various variables from stationary values, the required relation (4.14) follows the equation written above.

Expansion of the principle of invariance of correlations with respect to reversion of time for stationary situations allows proving the principle of symmetry of kinetic coefficients.

4.6 The simplest samples

In this section for illustration, we shall consider, following works (Pokrovskii 2005, 2013), some simple cases of thermodynamic system in stationary non-equilibrium situations. The values of characteristic functions of the system are constant, though there are changes of entropy from the outer fluxes and internal production of entropy within the system, which, correspondingly, can be written, according to equations (2.14) and (2.16), as

$$\frac{d_e\,S}{dt} = \frac{1}{T}\left(\frac{dQ}{dt} + \sum_\alpha \eta_\alpha \frac{dc_\alpha}{dt}\right), \quad \frac{d_iS}{dt} = -\frac{1}{T}\sum_k \frac{1}{\tau_k}\,\Xi_k\xi_k. \tag{4.18}$$

Let us recall that fluxes of heat and substances into the system are considered as constant; η_α is a contribution of energy into specific entropy connected with flux of substance α.

At given constant fluxes, the contents of the system are unchanged, albeit of exchange of heat and matter with surroundings, and one can consider the system to be in steady state. This case can be compared with the equilibrium situation and can be classified as a quasi-equilibrium situation.

4.6.1 Heat flux

When two systems with different temperatures are brought into contact, the flux of heat starts to flow from one system into another. The amount of heat ΔQ, which is received by one of the systems through its surface, initially is not distributed homogeneously, temperatures are different in different points of the system, and one needs in internal variables to characterize the state of the system.

Consider some substance in a small volume $\Delta x\ \Delta y\ \Delta z$ as a thermodynamic system with a temperature gradient $\nabla_x T$ and with a heat flow J along the x-axis

$$J = \frac{1}{\Delta y\ \Delta z}\frac{\Delta Q}{\Delta t}, \quad \nabla_x T = \frac{\Delta T}{\Delta x}. \tag{4.19}$$

The flow and gradient are always locally connected to each other at the point in the system under consideration. In the simplest case we have the local relation between the quantities

$$J = -\lambda\ \nabla_x T. \tag{4.20}$$

We consider the temperature gradient $\nabla_x T$ to be an internal variable. In this case, the average rates of thermal movement of molecules in different layers are different, and this can be registered in a different way.

If there are no external influences, the gradient of temperature is always tending to decrease, and change of gradient, if the above equation is valid, is determined by the following equation

$$\frac{\partial \nabla T}{\partial t} = \lambda \frac{\partial^2 \nabla T}{\partial x \, \partial x} \tag{4.21}$$

or in the other form, using the finite differences

$$\frac{\mathrm{d}\nabla T(x)}{\mathrm{d}t} = -\lambda(\nabla T(x - \Delta x) - 2\nabla T(x) + \nabla T(x + \Delta x)), \tag{4.22}$$

It means that relaxation of the temperature gradient in one point is not independent from values of temperature gradient in other points of the system. However, normal modes, which relax independently from each other, can be introduced in this case. We simplify the situation and consider a relaxation equation for a gradient of temperature in the form

$$\frac{\mathrm{d}\nabla T}{\mathrm{d}t} = -\frac{1}{\tau} \nabla T, \tag{4.23}$$

where τ is a time of relaxation of the temperature gradient which can be considered as an internal complexity variable. So, gradient of temperature $\nabla_x T$ can serve as an example of an internal variable.

We consider the steady-state situation, and the flux of heat defined by the relations (4.19) and (4.20), determine, according to the first equation from the relations (4.18), the flux of entropy

$$\frac{d_e S}{dt} = \frac{1}{\Delta x}\left(\frac{J}{T} - \frac{J}{T + \Delta T}\right) \approx \frac{J}{T^2} \nabla_x T. \tag{4.24}$$

However, in steady-state situation, the thermodynamic state of the volume does not change. So, internal production of entropy is

$$\frac{d_i S}{dt} = -\frac{d_e S}{dt} = \frac{\lambda}{T^2} (\nabla_x T)^2. \tag{4.25}$$

One can see that the gradient of temperature $\nabla_x T$ is the cause of the permanent internal production of entropy which, according to the second equation from the set (4.18), can also be presented as

$$\frac{d_i S}{dt} = \frac{1}{T} \frac{1}{\tau} \Xi \, \xi, \quad \xi = \nabla_x T \tag{4.26}$$

where τ is a time of relaxation of the internal variable, which is temperature gradient in this case. Now we can compare relation (4.25) and (4.26) to find the thermodynamic force

$$\Xi = \frac{\lambda\tau}{T}\,\xi, \qquad \xi = \nabla_x T.$$

Eventually, formula (2.11) can be used to calculate entropy of the system

$$S - S_0 = -\frac{1}{T}\int_0^{\nabla_x T} \Xi\, d\xi = -\frac{\lambda\tau}{2T^2}(\nabla_x T)^2. \tag{4.27}$$

4.6.2 Diffusion

Quite similar, one can consider a situation in a small volume $\Delta x\ \Delta y\ \Delta z$ with a permanent flux of substance

$$J = \frac{1}{\Delta y\ \Delta z}\,\frac{\mathrm{d}m}{\mathrm{d}t} = -D\nabla_x c \tag{4.28}$$

coming through along the x-axis due to the concentration gradient

$$\nabla_x c = \frac{\Delta c}{\Delta x}.$$

The flux of substances carries a flux of energy (chemical potential μ and specific entropy η) that depend on concentration

$$\eta = \eta(c), \qquad \frac{\partial\eta}{\partial c} \leqslant 0.$$

According to the first equation from the set (4.18), the flux of entropy is

$$\frac{\mathrm{d}_e S}{\mathrm{d}t} = \frac{J}{T}\,\frac{\eta(c + \Delta c) - \eta(c)}{\Delta x} = \frac{J}{T}\,\frac{\partial\eta}{\partial c}\nabla_x c,$$

or, taking equation (4.28) into account, the flux of entropy is

$$\frac{\mathrm{d}_e S}{\mathrm{d}t} = \frac{1}{\kappa T}\,(D\ \nabla_x c)^2, \qquad D = -\kappa\,\frac{\partial\eta}{\partial c}. \tag{4.29}$$

In steady-state situation, the outer flux is compensated by the internal production of entropy

$$\frac{\mathrm{d}_i S}{\mathrm{d}t} = -\frac{\mathrm{d}_e S}{\mathrm{d}t} = -\frac{1}{\kappa T}\,(D\ \nabla_x c)^2. \tag{4.30}$$

One can see that there is permanent internal production of entropy that is connected with gradient of concentration $\nabla_x c$. The further speculations are quite similar to those in the previous subsection, so that one can define the thermodynamic force

$$\Xi = \frac{\tau}{\kappa}\,(D)^2\,\xi, \qquad \xi = \nabla_x c$$

and eventually calculate the non-equilibrium contribution to entropy of the system

$$S - S_0 = -\frac{\tau}{2\kappa T}(D\,\nabla_x c)^2. \tag{4.31}$$

4.6.3 The couple fluxes

We continue to consider a portion of matter in a small volume $\Delta x\,\Delta y\,\Delta z$ as a thermodynamic system. In addition to the cases, when a stationary state is maintained by fluxes of heat and particles separately, let us consider the simultaneous action of these factors. We can assume that streams of heat and a substance are moving along the x-axis, so that

$$J_T = \frac{1}{\Delta y\,\Delta z}\frac{\Delta Q}{\Delta t}, \quad J_C = \frac{1}{\Delta y\,\Delta z}\frac{\Delta m}{\Delta t}.$$

The fluxes of heat and substance are contributing independently in the entropy of the volume. The flux of heat, according to equation (4.18), determines the change of density of entropy in the volume

$$\frac{d_{\text{heat}}S}{dt} = \frac{1}{\Delta x}\left(\frac{J_T}{T} - \frac{J_T}{T+\Delta T}\right) \approx \frac{J_T}{T^2}\nabla_x T. \tag{4.32}$$

In a similar way, one can define the change of density of entropy due to the flux of the substance

$$\frac{d_{\text{subs}}S}{dt} = \frac{J_C}{T}\frac{\eta(c+\Delta c)-\eta(c)}{\Delta x} \approx \frac{J_C}{T}\frac{\partial\eta}{\partial c}\nabla_x c. \tag{4.33}$$

The system is assumed to be characterized by specific entropy $\eta(c)$ that depends on concentration c.

It is convenient to introduce special symbols for the gradients of temperature and concentration, which can be considered to be internal complexity variables

$$\xi_T = \nabla_x T = \frac{\Delta T}{\Delta x} \quad \xi_C = \nabla_x c = \frac{\Delta c}{\Delta x}.$$

Thus, the change of density of entropy, due to the external fluxes of heat and substance, can be written as follows

$$\frac{d_e S}{dt} = \frac{J_T}{T^2}\xi_T + \frac{J_C}{T}\frac{\partial\eta}{\partial c}\xi_C. \tag{4.34}$$

The fluxes and gradients are connected with each other locally. It is well known (Prigogine 1961, Groot and Mazur 1962, Maugin and Muschik 1994, Kondepudi and Prigogine 1999, Jou *et al* 2001), as an empirical fact, that, at simultaneous presence of the thermal and diffusion gradients, ξ_T and ξ_C, cross effects occur: a gradient of temperature induces a flux of the substance and, vice versa, a gradient of concentration induces a flux of heat, so that, according to the last parity from equation (4.3), we record linear equations connecting the fluxes and gradients for the considered system

$$J_T = - R_{TT}\,\xi_T - R_{TC}\,\xi_C,$$
$$J_C = - R_{CT}\,\xi_T - R_{CC}\,\xi_C. \tag{4.35}$$

The fluxes of heat and matter determine the change in entropy (4.34), which is compensated by internal entropy production, represented by the equations (4.10), in the form of

$$\frac{d_iS}{dt} = \frac{1}{T}\,[(R_{TT}\,\xi_T + R_{TC}\,\xi_C)\,\Xi_T + (R_{CT}\,\xi_T + R_{CC}\,\xi_C)\,\Xi_C]. \tag{4.36}$$

In a steady-state situation, the thermodynamic state of the system does not change, so that, taking relation (4.34) into account, one has another expression for production of entropy

$$\frac{d_iS}{dt} = -\frac{J_T}{T^2}\,\xi_T - \frac{J_C}{T}\frac{\partial\eta}{\partial c}\xi_C. \tag{4.37}$$

Comparing equations (4.36) and (4.37), one finds the relations between fluxes and thermodynamic forces for the stationary case

$$J_T = - T\,(R_{TT}\,\Xi_T + R_{CT}\,\Xi_C),$$
$$J_C = -\left(\frac{\partial\eta}{\partial c}\right)^{-1}(R_{TC}\,\Xi_T + R_{CC}\,\Xi_C). \tag{4.38}$$

The condition of symmetry of kinetic coefficients (4.14) is followed by an equation

$$\frac{1}{T}\,R_{TC} = \frac{\partial\eta}{\partial c}\,R_{CT}. \tag{4.39}$$

The thermodynamic forces Ξ_T and Ξ_C, as functions of complexity variables, can be found from equations that are following (4.35) and (4.38)

$$R_{TT}\,\Xi_T + R_{CT}\,\Xi_C = \frac{1}{T}(R_{TT}\,\xi_T + R_{TC}\,\xi_C)$$
$$R_{TC}\,\Xi_T + R_{CC}\,\Xi_C = \frac{\partial\eta}{\partial c}\,(R_{CT}\,\xi_T + R_{CC}\,\xi_C). \tag{4.40}$$

Entropy of the system is defined by formula (2.11), which, in the considered case, takes the form

$$S - S_0 = -\frac{1}{T}\int_0^{\nabla_x T}\int_0^{\nabla_x c}(\Xi_T\,d\xi_T + \Xi_C\,d\xi_C). \tag{4.41}$$

After the calculation of the integrals, one gets a simple formula for entropy as a function of internal variables

$$S - S_0 = -\frac{1}{2T}\left[\frac{1}{T}(\nabla_x T)^2 + \frac{\partial \eta}{\partial c}(\nabla_x c)^2\right]. \tag{4.42}$$

It is remarkable, that the condition of integrability and existence of entropy of the system appears to be identical to Onsager's relation, that is equality of the non-diagonal components of a matrix in equations (4.38).

References

Groot de S R and Mazur P 1962 *Non-Equilibrium Thermodynamics* (Amsterdam: North-Holland)

Hackl K and Fischer F D 2008 On the relation between the principle of maximum dissipation and inelastic evolution given by dissipation potentials *Proc. R. Soc.* A **464** 117–32

Jou D, Casas-Vázquez J and Lebon G 2001 *Extended Irreversible Thermodynamics* 3rd edn (Berlin: Springer)

Kleidon A 2009 Nonequilibrium thermodynamics and maximum entropy production in the Earth system. Applications and implications *Naturwissenschaften* **96** 653–77

Kondepudi D and Prigogine I 1999 *Modern Thermodynamics: From Heat Engines to Dissipative Structures* (Chichester: Wiley)

Landau L D and Lifshitz E M 1986 *Statistical Physics* vol 1 3rd edn (Oxford: Pergamon)

Maugin G A and Muschik W 1994 Thermodynamics with internal variables part I. General concepts *J. Non-Equil. Thermodyn.* **19** 217–49

Pokrovskii V N 2005 Extended thermodynamics in a discrete-system approach *Eur. J. Phys.* **26** 769–81

Pokrovskii V N 2013 A Derivation of the main relations of non-equilibrium thermodynamics *ISRN Thermodyn.* **2013** 906136

Prigogine I 1961 *Introduction to Thermodynamics of Irreversible Processes* 2nd edn (New York: Interscience)

IOP Publishing

Thermodynamics of Complex Systems
Principles and applications
Vladimir N Pokrovskii

Chapter 5

Statistical interpretation of thermodynamic characteristics

In the previous chapters, the thermodynamic characteristics were introduced and considered irrespective of any assumptions about the detailed composition of the system, which, from the other point of view, can be looked at as at the system consisting of many particles. One can expect that this would allow one to develop methods of calculation of the macroscopic quantities via microscopic characteristics of the system. The problem was being solved by statistical mechanics, so that one can consider that thermodynamics and statistical mechanics are inseparable, while the latter provides the foundation of the former. Nevertheless the thermodynamics is formulated independently; the disciplines have the same object, but different methods, and, in fact, the principles of thermodynamics appear to be primary in comparison with the principles of statistical physics.

5.1 Two approaches to the thermodynamic system

5.1.1 Macroscopic description

In the first and second chapters, it has been shown, that it is possible to characterize a thermodynamic system, in both equilibrium, and non-equilibrium states, with macroscopic quantities, among them internal total energy U, free energy F and entropy S

$$\begin{aligned} U(S, \boldsymbol{x}, \boldsymbol{\xi}) &= U(S, x_1, x_2, \ldots, x_n;\ \xi_1, \xi_2, \ldots), \\ F(T, \boldsymbol{x}, \boldsymbol{\xi}) &= F(T, x_1, x_2, \ldots, x_n;\ \xi_1, \xi_2, \ldots), \\ S(T, \boldsymbol{x}, \boldsymbol{\xi}) &= S(T, x_1, x_2, \ldots, x_n;\ \xi_1, \xi_2, \ldots). \end{aligned} \tag{5.1}$$

The quantities are considered as functions of variables of state: temperatures T, fundamental variables $x_1, x_2, \ldots, x_n$ and uncertain number of internal variables $\xi_1, \xi_2, \ldots$.

doi:10.1088/978-0-7503-3451-8ch5

By way of introduction, the thermodynamic quantities, in both equilibrium, and non-equilibrium states, are connected among themselves by universal parities

$$U = F + TS, \qquad S = -\frac{\partial F}{\partial T}, \qquad T = \frac{\partial U}{\partial S}. \tag{5.2}$$

Thermodynamic functions are introduced phenomenologically, on the basis of fundamental principles, and the relations among the functions do not depend on any interpretation or details of a structure of system.

5.1.2 Microscopic description

From the other side, the thermodynamic system is a system of many particles, so it is possible to consider the system as a dynamic system of interacting particles—the microscopic description. From the point of view of classical mechanics (see, for example, Landau and Lifshitz 1969), the state of each particle of the system is set by coordinate and momentum, so that an instant state of the system of N particles is fixed by the set of coordinates $q_1, q_2, \dots q_{3N}$ and momenta $p_1, p_2, \dots p_{3N}$. One needs in $6N$ variables in total to fix a state of the system, it is possible to tell, a microstate of the system, which can be represented by a point in phase space. The coordinates and momenta of the particles of the system change according to the equations of motion

$$\frac{\mathrm{d}q_j}{\mathrm{d}t} = \frac{\partial H}{\partial p_j}, \qquad \frac{\mathrm{d}p_j}{\mathrm{d}t} = -\frac{\partial H}{\partial q_j}, \qquad j = 1, 2, \dots 3N, \tag{5.3}$$

where H is a function of Hamilton of the considered system of particles. The function of Hamilton represents the expression for the total energy of the system, recorded via coordinates and momenta of all particles. In the case, for example, of a simple system, consisting of N identical interacting particles, the function of Hamilton looks like

$$H(p_1, p_2, \dots p_{3N}; q_1, q_2, \dots q_{3N}) = \sum_{j=1}^{3N} \frac{p_j^2}{2m} + V(q_1, q_2, \dots q_{3N}), \tag{5.4}$$

Notice, that the microscopic description in the Hamiltonian form, also as well as the macroscopic description, has universal character: peculiarities of behaviour of the system are included in the function of Hamilton.

5.1.3 How to compare two approaches?

The macroscopic and microscopic descriptions must correspond to each other: one can imagine that there is a way to transition from the microscopic variables (the coordinates and momenta of all particles) to macroscopic ones; it is necessary only to find such a way. It is possible to believe, that temperature T and thermodynamic variables $x_1, x_2, \dots, x_n; \xi_1, \xi_2, \dots$ are macroscopic analogues of coordinates

$q_1, q_2, \ldots q_{3N}$ and momenta $p_1, p_2, \ldots p_{3N}$ of the same system considered from the other point of view.

It is necessary to consider the thermodynamic functions as some fixed combination of microscopic variables, anyway, it is possible to note, that thermodynamic total energy of the system (in a certain point in time) corresponds to the function of Hamilton, and we can record

$$U(S, \boldsymbol{x}, \boldsymbol{\xi}) = H(p_1, p_2, \ldots p_{3N}; q_1, q_2, \ldots q_{3N}; \boldsymbol{x}). \tag{5.5}$$

The function of Hamilton includes, generally, one or several controllable parameters $\boldsymbol{x}$, corresponding to thermodynamic fundamental variables. At differentiation of the function and taking into account parities (5.3), we find that variation of total energy of the system is equal to work connected with supervising parameters (fundamental variables of thermodynamic system)

$$\frac{\mathrm{d}H(\boldsymbol{q}, \boldsymbol{p}, \boldsymbol{x})}{\mathrm{d}t} = \sum_j \frac{\partial H}{\partial x_j}\frac{\mathrm{d}x_j}{\mathrm{d}t} = -\sum_j X_j(\boldsymbol{q}, \boldsymbol{p}, \boldsymbol{x})\frac{\mathrm{d}x_j}{\mathrm{d}t},$$

where X_j is work of the system on the environment.

For the complete association of two approaches, the arguments of the thermodynamic total energy: entropy S and internal variables $\boldsymbol{\xi}$ should be determined as functionals of the phase variables: coordinates $q_1, q_2, \ldots q_{3N}$ and momenta $p_1, p_2, \ldots p_{3N}$. The problem of definition of rules of combinations and transformations of the phase variable (coordinates and impulses of all particles), which allows definition of all thermodynamic quantities, is considered by statistical physics.

Because of the great number of particles in the thermodynamic system, there is no hope to present a complete description of the microscopic dynamics, but the microscopic interpretation of the system allows one to introduce a statistical ensemble of realizations of a macroscopic state of the system, which determines a function of distribution of microscopic states of the system. For the thermodynamic systems that are in equilibrium states, rules of construction of the statistical ensemble, allowing calculation of average values of various thermodynamic quantities, have been developed by Gibbs (1902), and it is his greatest achievement. The application to the cases of systems in equilibrium proceeded rapidly and with exceptional success: equilibrium statistical mechanics gave—starting from the microscopic level—foundations to thermodynamics in equilibrium. Applications to non-equilibrium systems began, mainly, with the case of local equilibrium in the linear approximation.

5.2 Introduction of statistical ensemble

Let us consider a thermodynamic system that is in equilibrium or in one of the steady states, so that its macroscopic state is fixed with values of thermodynamic variables T, $\boldsymbol{x}$, $\boldsymbol{\xi}$ and does not change during time. Characteristic functions of the system: internal energy U, free energy F and entropy S have constant values. As a set of particles, the system obeys the laws of classical mechanics and, if the system is isolated, has a certain constant value of total energy U. If from the thermodynamic point of view the system is being in the equilibrium state ($\boldsymbol{\xi} = 0$), the value of total energy U coincides with the value of internal thermal energy E. Otherwise $U \geqslant E$.

The complete description of the system of many particles assumes that behaviour of each particle is considered. A momentary state of the system is described by the set of coordinates $q_1, q_2, \ldots q_{3N}$ and momenta $p_1, p_2, \ldots p_{3N}$, so that one needs $6N$ variables to specify the state of the system, one can say microstate of the system. If initial values of coordinate and momenta of a particle are specified (the point in phase space) a trajectory of each particle, according to the equations recorded above (5.3), is determined unequivocally

$$q_j = q_j(\boldsymbol{q}(0), \boldsymbol{p}(0), t), \quad p_j = p_j(\boldsymbol{q}(0), \boldsymbol{p}(0), t), \quad j = 1, 2, \ldots 3N.$$

The law of development of phase variables determines a trajectory in a $6N$-dimensional phase space. The trajectory of development of the system runs through many points in phase space, each of which represents the thermodynamic system in the preset macrostate. Differently, the chosen macrostate of the system can be realized by many various ways. It also means, that one macrostate corresponds to a set of microrealizations (microstates).

The macroscopic properties of the system at constant values of macroscopic variables U, $\boldsymbol{x}$, $\boldsymbol{\xi}$ are invariant in relation to the shift over time, and at consecutive shifts with an interval of time Δt during a sufficiently long period, it is possible to prepare a set of copies of the system corresponding to one macroscopic state, fixed by values of the variables U, $\boldsymbol{x}$, $\boldsymbol{\xi}$. Each copy of the system is fixed by the certain set of values of the coordinates and momenta $\boldsymbol{q}$ and $\boldsymbol{p}$, and it is possible to speak about the copy as about a microstate of the thermodynamic system. Thus, the set of copies of the system can be established. The number of copies $\Omega(U, \boldsymbol{x}, \boldsymbol{\xi})$ can be named also as the number of microstates of the system at the preset values U, $\boldsymbol{x}$, $\boldsymbol{\xi}$. In each point in time the system is in one of the microstates, but we do not know in which.

It is possible to assume, that the system can equiprobably be in any of its available microstates, and this statement is valid, apparently, for both the equilibrium state of the system, and for a stationary non-equilibrium situation. The statement is considered as *the basic postulate of statistical thermodynamics*.

It is easy to see, that the requirement of a constancy of the preset thermodynamic characteristics defines some restrictions on values of coordinates and momenta of particles. For example, the coordinates of particles of the system that is restricted in space accept values in the certain area. If total energy of the isolated system is constant a condition

$$U = \sum_j \frac{p_j^2}{2m} + V(q_1, q_2, \ldots q_{3N})$$

defines some restrictions on values of coordinates and momenta. The requirement of a constancy of internal variables also establishes some restrictions.

To describe population of the phase space with the created copies, we shall assume, that the phase space of the system can be divided into cells with volume

$$\prod_{\alpha=1}^{3N} \Delta q_\alpha \Delta p_\alpha .$$

Due to the restrictions established by values of macroscopic variables U, $\boldsymbol{x}$, $\boldsymbol{\xi}$, some cells of the phase space appear inaccessible: the cells which contain a part of a trajectory of the system are allocated. Not all cells from the available cells of the phase space appear equally occupied. The number of copies $\Omega_{q,p} = \Omega_{q,p}(U, \boldsymbol{x}, \boldsymbol{\xi})$ in the cell with the certain volume and with specific values of coordinates $\boldsymbol{q}$, $\boldsymbol{p}$ is, apparently, a part of the total of available cells, so that it is possible to record

$$\sum_{q,p} \Omega_{q,p}(U, \boldsymbol{x}, \boldsymbol{\xi}) = \Omega(U, \boldsymbol{x}, \boldsymbol{\xi}). \tag{5.6}$$

The summation is carried out over all cells of phase space.

The total number of microstates of the system $\Omega(U, \boldsymbol{x}, \boldsymbol{\xi})$ depends on the arbitrary interval of time of the shift used for construction of the ensemble. Apart from that, the number of microstates of the system in a cell with preset values of coordinates $\Omega_{q,p}(U, \boldsymbol{x}, \boldsymbol{\xi})$ depends on the fixed size of the cell in the phase space[1]. To exclude dependence on the arbitrary size, consider, firstly, the ratio of the quantities that determine the probability of distribution of copies over the cells of the phase space

$$P_{U,x,\xi}(\boldsymbol{q}, \boldsymbol{p}) = \frac{\Omega_{q,p}(U, \boldsymbol{x}, \boldsymbol{\xi})}{\Omega(U, \boldsymbol{x}, \boldsymbol{\xi})}. \tag{5.7}$$

This quantity depends, apparently, on the volume of the elementary cell. To introduce a characteristic that is independent on the volume of the cell, it is convenient to introduce the density of distribution of probability $\rho(\boldsymbol{q}, \boldsymbol{p})$, defining the probability to find the system in the element of phase space with the coordinates $\boldsymbol{q}$ and $\boldsymbol{p}$

$$P(\boldsymbol{q}, \boldsymbol{p}) = \rho(\boldsymbol{q}, \boldsymbol{p}) \Delta \boldsymbol{q} \Delta \boldsymbol{p}. \tag{5.8}$$

The distribution function is considered to be normalized

$$\int \rho(\boldsymbol{q}, \boldsymbol{p})\, \mathrm{d}\boldsymbol{q} \mathrm{d}\boldsymbol{p} = 1. \tag{5.9}$$

[1] In classic physics the size of the cell is introduced artificially, but in quantum mechanics there is a natural scale: the size of the cell is limited by the Planck constant

$$\prod_{j=1}^{3N} \Delta q_j \Delta p_j > \hbar^{3N}.$$

Thus, the statistical ensemble can be built, if the macroscopic characteristics of the system are unchangeable during some interval of time, which means that only equilibrium or stationary non-equilibrium states of a thermodynamic system can be considered by this method. This procedure allows one to determine the distribution function that does not depend explicitly on time. If the distribution function is known, the mean values of any function of coordinates and momenta can be directly calculated.

5.3 Statistics of an isolated system

5.3.1 Microcanonic distribution

One of the main characteristics of the system of particles is energy, which is a function of phase variables and has a certain value at any moment. Energy of the isolated system has a certain constant value U. If the system is immobile and isolated, one can take into account the laws of conservations of momentum and total energy of the system U, so that the trajectory is situated on the surface

$$\sum_j^{3N} q_j = 0, \quad \sum_j^{3N} p_j = 0, \quad \sum_j^{3N} \frac{p_j^2}{2m} + V(q_1, q_2, \dots q_{3N}) = U.$$

A statistical ensemble for an isolated system can be easily built for the equilibrium situation. Total energy of an isolated system coincides with internal thermal energy and does not depend on time, so that the distribution function ought to be written as

$$\rho(\boldsymbol{q}, \boldsymbol{p}) = \delta(U - H(\boldsymbol{q}, \boldsymbol{p})) \tag{5.10}$$

This distribution function is called microcanonical distribution.

The isolated system can be in a non-equilibrium situation, though it is hard to imagine the non-equilibrium to be stationary; the internal variables arising during influence on the system, disappear eventually. In the case where it would be possible to present system in a stationary non-equilibrium situation and have an expression for the internal variable through coordinates and momenta, the distribution function could be recorded in the form

$$\rho(\boldsymbol{q}, \boldsymbol{p}) = \delta(U - H(\boldsymbol{q}, \boldsymbol{p}))\delta(\xi - \xi(\boldsymbol{q}, \boldsymbol{p})).$$

In this case total energy of the isolated system U does not coincide with internal thermal energy E, $U \geqslant E$.

5.3.2 Entropy of an isolated system

One of the main characteristics of the thermodynamic system is entropy S, about which it is known that it has the greatest value in the equilibrium state. To find an expression for entropy, consider an isolate system, consisting of two isolated subsystems, and use the property of additivity of entropy

$$S = S_1 + S_2.$$

We assume, that systems are in the equilibrium situation.

At given values of macroscopic parameters, the microscopic situation of each of the systems is characterized by the numbers of acceptable microstates

$$\Omega_1 = \Omega_1(N_1, E_1), \quad \Omega_2 = \Omega_2(N_2, E_2).$$

The number of microstates of the combined system $\Omega = \Omega_{1+2}$ depends, apparently, on conditions of association of two systems. If systems are isolated from each other, the total number of microstates is defined by the relation

$$\Omega = \Omega(N_1, N_2, E_1, E_2) = \Omega_1(N_1, E_1)\Omega_2(N_2, E_2). \tag{5.11}$$

Of course, in the cases, when the systems are not isolated, that is, an exchange of particles and energy is possible, one has the other relation

$$\Omega = \Omega(N_1 + N_2, E_1 + E_2) > \Omega_1(N_1, E_1)\Omega_2(N_2, E_2). \tag{5.12}$$

Apparently, it was Boltzmann (Sommerfeld 1952), who first arrived at the idea to compare entropy with the number of microstates of the system and considered the dependence

$$S = S(\Omega).$$

Considering, that each subsystem is a big system, so that the interaction between the systems can be neglected and can be accepted parity (5.11) for the combined system, by virtue of the additivity of entropy, the relation can be recorded

$$S(\Omega) = S(\Omega_1\Omega_2) = S(\Omega_1) + S(\Omega_2).$$

Having differentiated this equation according to the each of the variables, one gets

$$\Omega_2\frac{\partial S(\Omega)}{\partial \Omega} = \frac{\partial S(\Omega_1)}{\partial \Omega_1}, \quad \Omega_1\frac{\partial S(\Omega)}{\partial \Omega} = \frac{\partial S(\Omega_2)}{\partial \Omega_2}.$$

This equations allow us to write

$$\Omega_1\frac{\partial S(\Omega_1)}{\partial \Omega_1} = \Omega_2\frac{\partial S(\Omega_2)}{\partial \Omega_2} = \Omega\frac{\partial S(\Omega)}{\partial \Omega} = k,$$

where k is a constant, so that for any isolated system one has

$$S(\Omega) = k \ln \Omega. \tag{5.13}$$

By virtue of equation (5.2), the temperature can be calculated

$$\frac{1}{T} = \frac{\partial S}{\partial U} = \frac{\partial \ln \Omega}{\partial U}. \tag{5.14}$$

It is convenient to consider entropy as non-dimensional quantity and to assume in equation (5.13) $k = 1$. In this case, temperature is measured in energy units.

5.4 Statistics of closed system

The closed system has invariable composition, but energy of the system is not constant; in the thermal form energy is passed from the system to the environment and vice versa. The distribution function of the system in equilibrium in the general form can be found according to the Gibbs method (Gibbs 1902). The found distribution function allows us to specify universal expressions for thermodynamic functions.

5.4.1 Canonical distribution

To approach the problem, one can consider an isolated system consisting of two thermodynamic systems, while one of them is situated inside the other system, which is large as compared with the other system and is called thermostat, as is depicted in figure 5.1. The combined system is isolated and considered to be in equilibrium at temperature T, whereas the exact value energy of the entire system can be written as the sum

$$E^0 = E + E_T = \text{const}.$$

There is exchange of heat between the small system and thermostat, so that the smaller system is classified as a closed system.

According to equation (5.7), the probability for the small system to have an exact energy value E can be written as

$$P(E) = \frac{\Omega(E)\,\Omega_T(E_T)}{\Omega^0(E^0)}, \tag{5.15}$$

where $\Omega(E)$ and $\Omega_T(E_T)$ are numbers of microstates of the small system and thermostat with the fixed values of energy E and E_T, correspondingly. The number of microstates $\Omega^0(E^0)$ of the whole system is the product of the numbers of microstates of the subsystems, if the subsystems have fixed values of energy, as shown by equation (5.11). If the values of energy of the subsystems are not fixed (at any internal realizations of the systems), this number is larger than the product of microstate numbers of the subsystems, that is, according to equation (5.12),

$$\Omega^0(E^0) > \Omega(E)\,\Omega_t(E_t).$$

All the microrealisations of the system are considered as equivalent at a given value of energy, so that the probability for the small system to have the energy value E and to be in a cell with particular values of coordinates and momentum $\boldsymbol{q}$, $\boldsymbol{p}$ is

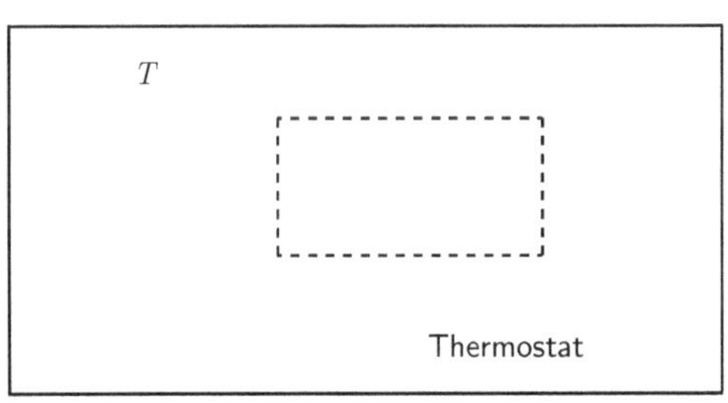

Figure 5.1. The scheme of the situation.

$$P_E(\boldsymbol{q},\boldsymbol{p}) = \frac{1}{\Omega(E)} = \frac{\Omega_T(E_T)}{\Omega^0(E^0)}. \tag{5.16}$$

It is assumed that there is only one microrealisation, belonging to the cell with the index q, p, that is, the microstate is non-degenerate.

The fixed value of energy of a subsystem E is connected with specific values of coordinates and momentum $\boldsymbol{q}$, $\boldsymbol{p}$ through the Hamilton function (5.4), so that the above recorded parity defines equation

$$P_E(\boldsymbol{q},\boldsymbol{p}) = \frac{\Omega_T[E^0 - H(\boldsymbol{q},\boldsymbol{p})]}{\Omega^0(E^0)}.$$

One can use the assumption that the considered system is small as compared with the thermostat, which means $E \ll E_T$, and one can transform the relation

$$\ln \Omega_T[E^0 - H(\boldsymbol{q},\boldsymbol{p})] = \ln \Omega_T(E^0) - \left(\frac{\partial \ln \Omega_T(E_T)}{\partial E_T}\right)_{E_T=E^0} H(\boldsymbol{q},\boldsymbol{p}) + \cdots.$$

In virtue of one of the above relations (5.14)

$$\left(\frac{\partial \ln \Omega_T(E_T)}{\partial E_T}\right)_{E_T=E^0} = \frac{1}{kT},$$

so that

$$\Omega_T(E_T) = \Omega_T(E^0)\exp\left(-\frac{H(\boldsymbol{q},\boldsymbol{p})}{kT}\right) + \cdots.$$

Then, one can return to the expression for the probability (5.16). Taking into account that for the large thermostat $\Omega_T(E^0) \approx \Omega^0(E^0)$, one has for the probability that the small system has energy value E and is in non-degenerated microstate labeled $\boldsymbol{q}$, $\boldsymbol{p}$

$$P_E(\boldsymbol{q},\boldsymbol{p}) = \exp\left(-\frac{H(\boldsymbol{q},\boldsymbol{p})}{kT}\right). \tag{5.17}$$

Now we can come to the function of probability density

$$P_E(\boldsymbol{q},\boldsymbol{p}) = \rho[H(\boldsymbol{q},\boldsymbol{p})]\Delta\boldsymbol{q}\Delta\boldsymbol{p}. \tag{5.18}$$

One can write the normalized function of distribution of the probability

$$\begin{aligned} \rho(\boldsymbol{q},\boldsymbol{p}) &= \frac{1}{Z(T)}\exp\left(-\frac{H(\boldsymbol{q},\boldsymbol{p})}{kT}\right), \\ Z(T) &= \int \exp\left(-\frac{H(\boldsymbol{q},\boldsymbol{p})}{kT}\right)\mathrm{d}\boldsymbol{q}\mathrm{d}\boldsymbol{p}. \end{aligned} \tag{5.19}$$

This is the canonical distribution for a closed system in the equilibrium situation.

The function can be also written as a distribution over values of energy

$$\rho(E) = \frac{1}{Z(E,\, T)} \exp\left(-\frac{E}{kT}\right),$$
$$Z(E,\, T) = \int_{E=\text{const.}} \exp\left(-\frac{H(\boldsymbol{q},\, \boldsymbol{p})}{kT}\right) \mathrm{d}\boldsymbol{q}\mathrm{d}\boldsymbol{p}. \tag{5.20}$$

5.4.2 Thermodynamic characteristics

The probability function allows one to calculate thermodynamic function for the equilibrium situation as mean values of corresponding quantities and find important relations.

Internal thermal energy E for an equilibrium state coincides with total internal energy of system U and is, apparently, the mean value of energy as a function of coordinates and momentum

$$E(t) = \int H(\boldsymbol{q},\, \boldsymbol{p})\, \rho(\boldsymbol{q},\, \boldsymbol{p})\, \mathrm{d}\boldsymbol{q}\mathrm{d}\boldsymbol{p}. \tag{5.21}$$

According to equation (5.13), one can calculate entropy as mean value of the logarithm of number of acceptable states of the system at fixed value of energy, that is

$$S = -\sum_E P(E) \ln \frac{1}{\Omega(E)}.$$

In virtue of equations (5.18), the expression for entropy of thermodynamic system can be rewritten as

$$S = -\sum P_E(\boldsymbol{q},\, \boldsymbol{p}) \ln P_E(\boldsymbol{q},\, \boldsymbol{p}) = -\int \ln \rho(\boldsymbol{q},\, \boldsymbol{p})\, \rho(\boldsymbol{q},\, \boldsymbol{p})\, \mathrm{d}\boldsymbol{q}\mathrm{d}\boldsymbol{p}. \tag{5.22}$$

One can use the specific form (5.19) of the distribution function to obtain the relation

$$\int \ln \rho(\boldsymbol{q},\, \boldsymbol{p})\, \rho(\boldsymbol{q},\, \boldsymbol{p})\, \mathrm{d}\boldsymbol{q}\mathrm{d}\boldsymbol{p} = -\ln Z(T) - \frac{1}{kT} \int H(\boldsymbol{q},\, \boldsymbol{p})\, \rho(\boldsymbol{q},\, \boldsymbol{p})\, \mathrm{d}\boldsymbol{q}\mathrm{d}\boldsymbol{p}. \tag{5.23}$$

In virtue of the above definitions of energy and entropy, one has the relation

$$E - TS = -kT \ln Z,$$

which allows one, comparing with thermodynamic relation (5.2), to define the expression for free energy

$$F = -kT \ln Z. \tag{5.24}$$

This definition, due to known thermodynamic relation (5.2), allows one to reduce the calculation all mean thermodynamic properties of a closed system to calculation of statistical integral $Z(T)$

$$S = -\left(\frac{\partial F}{\partial T}\right)_V = \ln Z + T\frac{\partial \ln Z}{\partial T}$$
$$p = -\left(\frac{\partial F}{\partial V}\right)_T = T\frac{\partial \ln Z}{\partial V}$$
$$E = F + TS = T^2\frac{\partial \ln Z}{\partial T}$$
$$\mu = \left(\frac{\partial F}{\partial N}\right)_{T,V} = -T\frac{\partial \ln Z}{\partial N}. \tag{5.25}$$

This reduction unifies the problem, but, nevertheless, calculation of the statistical integral for particular systems appears to be a difficult problem. This scheme was used for the calculation of thermodynamics characteristics of various objects (see, for example, Landau and Lifshitz 1986).

5.5 The principle of entropy maximum

It is known (see the discussion in the first and fourth chapters, sections 1.5 and 4.4) that equilibrium and stationary non-equilibrium states of a thermodynamic system appear to be specific states: entropy of the system in these states appears to have the maximal value in relation to variation of uncontrollable internal variables of the system. This statement can be considered as a principle—a heuristic principle of maximality of entropy. Having a presentation of entropy through a local distribution function, it is possible to state, that in relation to any deviation of function of distribution from the form corresponding to equilibrium or stationary non-equilibrium state, value of entropy decreases. It allows one to develop a variational method of evaluation of the distribution function, comprehensible for both systems in equilibrium, and systems being in stationary non-equilibrium states[2]. In the basis of this method, the definition of entropy of the system through the distribution function lies. The method was used to find a non-equilibrium distribution function, as will be illustrated further. Then, the open systems, when the number of particles in the system is not fixed, can be considered.

5.5.1 Classical functional

Under some assumption, it was found that the expression for entropy of the closed system can be written in the form

$$S = -\int \rho(\boldsymbol{q}, \boldsymbol{p})\ln \rho(\boldsymbol{q}, \boldsymbol{p})\mathrm{d}\boldsymbol{q}\mathrm{d}\boldsymbol{p}. \tag{5.26}$$

[2] The theory was developed by Jaynes (1957), referring to the theory of information, and was interpreted, as a method of maximization of information uncertainty or information entropy. The theory has also received the name *prognostic statistical mechanics*. Despite different names, it should be noticed that this is not an alternative approach or new paradigm in statistical physics, but a quite useful and practical variational method which unifies the derivation of distributions of probability for various situations.

The result concerns isolated or closed thermodynamic systems in equilibrium situations. We believe, that the expression for entropy of any thermodynamic system (including the open system, in an equilibrium or stationary non-equilibrium state) is formulated in the same way. In this section we consider the closed system with a certain number of particles in a stationary non-equilibrium situation. In this case it is convenient to consider local distribution function, which depends also on the space coordinate $\boldsymbol{x}$, but not on time $\rho(\boldsymbol{q}, \boldsymbol{p}, \boldsymbol{x})$, and one can look for the distribution function from the requirement of maximum of the above function at some extra requirements.

5.5.1.1 The restraints in a problem of maximization

To find a distribution function, we look for a point of an extremum of entropy, determined by expression (5.26), with some additional requirements. It is convenient to believe, that the function of distribution $\rho(\boldsymbol{q}, \boldsymbol{p})$, is normalized to unity

$$\int \rho(\boldsymbol{q}, \boldsymbol{p})\, \mathrm{d}\boldsymbol{q}\mathrm{d}\boldsymbol{p} = 1. \tag{5.27}$$

The condition of a constancy of the mean value of total energy is imposed on a closed system

$$\int H(\boldsymbol{q}, \boldsymbol{p})\rho(\boldsymbol{q}, \boldsymbol{p})\, \mathrm{d}\boldsymbol{q}\mathrm{d}\boldsymbol{p} = U, \tag{5.28}$$

where $H(\boldsymbol{q}, \boldsymbol{p})$ is the function of Hamilton defined by equation (5.4).

Apart from the above recorded parities, the distribution function of a non-equilibrium system should submit to some other restrictions, which should be expressed through some combinations of microscopic variables. For example, considering a gradient of temperature as an internal variable of a complex system, it is necessary to define this quantity in terms of microscopic variables. For example, for ideal gas, at the account of a parity $E = 3NT/2$, the average temperature in a point $\boldsymbol{x}$ is defined as

$$T(\boldsymbol{x}) = \frac{2}{3N} \int H(\boldsymbol{q}, \boldsymbol{p}, \boldsymbol{x})\rho(\boldsymbol{q}, \boldsymbol{p})\, \mathrm{d}\boldsymbol{q}\mathrm{d}\boldsymbol{p}.$$

Integration over momenta is fulfilled without restrictions, and integration on each of the spatial coordinates is carried out within the limits of small pieces about value $\boldsymbol{x}$. Differentiation of the written parity defines a gradient of temperature

$$\frac{\partial T}{\partial x_j} = \frac{2}{3N} \int \frac{\partial H(\boldsymbol{q}, \boldsymbol{p}, \boldsymbol{x})}{\partial x_j} \rho(\boldsymbol{q}, \boldsymbol{p})\, \mathrm{d}\boldsymbol{q}\mathrm{d}\boldsymbol{p}.$$

Probably, the greatest difficulty at the consideration of stationary non-equilibrium processes appears to express quantities of internal complexity variables via the microscopic variables. Assuming that it is possible to solve this problem for all quantities that are interesting for us, we record at the given values of internal variables the restrictions

$$\int B_\alpha(\boldsymbol{q}, \boldsymbol{p})\rho(\boldsymbol{q}, \boldsymbol{p})\, \mathrm{d}\boldsymbol{q}\mathrm{d}\boldsymbol{p} = \xi_\alpha, \quad \alpha = 1, 2, \ldots, s. \tag{5.29}$$

5.5.1.2 Distribution function

We address now the case of a complex system, when values of some internal variables are distinct from zero and are fixed. We assume, that the distribution function is defined by an extremum of function (5.26) with restrictions (5.27)–(5.29). We use the Lagrange method and, instead of a conditional extremum of function (5.26), look for an unconditional extremum of the criterion function

$$\begin{aligned}\Phi = &- \int \rho(\boldsymbol{q}, \boldsymbol{p})\ln \rho(\boldsymbol{q}, \boldsymbol{p})\mathrm{d}\boldsymbol{q}\mathrm{d}\boldsymbol{p} - \lambda_0 \int \rho(\boldsymbol{q}, \boldsymbol{p})\, \mathrm{d}\boldsymbol{q}\mathrm{d}\boldsymbol{p} \\ &- \beta \int H(\boldsymbol{q}, \boldsymbol{p})\rho(\boldsymbol{q}, \boldsymbol{p})\, \mathrm{d}\boldsymbol{q}\mathrm{d}\boldsymbol{p} \\ &- \sum_\alpha \lambda_\alpha \int B_\alpha(\boldsymbol{q}, \boldsymbol{p})\rho(\boldsymbol{q}, \boldsymbol{p})\, \mathrm{d}\boldsymbol{q}\mathrm{d}\boldsymbol{p},\end{aligned} \tag{5.30}$$

where λ_0, β and λ_α are the Lagrange multipliers.

The condition of an extremum defines function of distribution in the form of

$$\rho(\boldsymbol{q}, \boldsymbol{p}) = \exp\left(-1 - \lambda_0 - \beta H(\boldsymbol{q}, \boldsymbol{p}) - \sum_\alpha \lambda_\alpha B_\alpha(\boldsymbol{q}, \boldsymbol{p})\right).$$

The requirement of normalisation determines one of the multipliers

$$\exp(1 + \lambda_0) = \int \exp\left(-\beta H(\boldsymbol{q}, \boldsymbol{p}) - \sum_\alpha \lambda_\alpha B_\alpha(\boldsymbol{q}, \boldsymbol{p})\right) \mathrm{d}\boldsymbol{q}\mathrm{d}\boldsymbol{p},$$

so that the distribution function can be written in the form

$$\rho(\boldsymbol{q}, \boldsymbol{p}) = \exp\left(\Gamma(\beta, \lambda_1, \lambda_2, \ldots, \lambda_s) - \beta H(\boldsymbol{q}, \boldsymbol{p}) - \sum_\alpha \lambda_\alpha B_\alpha(\boldsymbol{q}, \boldsymbol{p})\right), \tag{5.31}$$

where

$$\Gamma(\beta, \lambda_1, \lambda_2, \ldots, \lambda_s) = -\ln \int \exp\left(-\beta H(\boldsymbol{q}, \boldsymbol{p}) - \sum_\alpha \lambda_\alpha B_\alpha(\boldsymbol{q}, \boldsymbol{p})\right) \mathrm{d}\boldsymbol{q}\mathrm{d}\boldsymbol{p}. \tag{5.32}$$

This quantity allows one to calculate the mean values of total energy and variables of complexity

$$\frac{\partial \Gamma}{\partial \beta} = U, \quad \frac{\partial \Gamma}{\partial \lambda_\alpha} = \xi_\alpha. \tag{5.33}$$

These relations allow one to determine the multipliers as a function of energy and complexity variables

$$\beta = \beta(U, \xi_1, \xi_2, \ldots, \xi_s), \qquad \lambda_\alpha = \lambda_\alpha(U, \xi_1, \xi_2, \ldots, \xi_s). \tag{5.34}$$

5.5.1.3 Thermodynamic characteristics

Addressing now the derivation of thermodynamic characteristics of the system, we shall note, first of all, that the mean value of the function of Hamilton is the thermodynamic total energy of the system

$$U = \int H(\boldsymbol{q}, \boldsymbol{p})\, \rho(\boldsymbol{q}, \boldsymbol{p})\, \mathrm{d}\boldsymbol{q}\mathrm{d}\boldsymbol{p}. \tag{5.35}$$

The mean value of logarithm of the distribution function

$$\ln \rho(\boldsymbol{q}, \boldsymbol{p}) = \Gamma(\beta, \lambda_1, \lambda_2, \ldots, \lambda_s) - \beta H(\boldsymbol{q}, \boldsymbol{p}) - \sum_\alpha \lambda_\alpha B_\alpha(\boldsymbol{q}, \boldsymbol{p})$$

defines entropy

$$S = -\Gamma(\beta, \lambda_1, \lambda_2, \ldots, \lambda_s) + \beta U + \sum_\alpha \lambda_\alpha \xi_\alpha. \tag{5.36}$$

The comparison with thermodynamic relation (5.2) allows one to define free energy and temperature of the system

$$F = T\left(\Gamma(\beta, \lambda_1, \lambda_2, \ldots, \lambda_s) - \sum_\alpha \lambda_\alpha \xi_\alpha\right), \qquad T = \frac{1}{\beta}.$$

The thermodynamic free energy is a function of temperature and complexity variables, and, using the relations (5.33), one can write the total differential of the quantity

$$\mathrm{d}F = \left(\Gamma(\beta, \lambda_1, \lambda_2, \ldots, \lambda_s) - \sum_\alpha \lambda_\alpha \xi_\alpha - \frac{U}{T}\right)\mathrm{d}T + T\sum_\alpha \lambda_\alpha \mathrm{d}\xi_\alpha.$$

In accordance with thermodynamic relation (5.2), the expression (with minus sign) inside the large round brackets coincides with entropy (5.36). The quantities $T\lambda_1, T\lambda_2, \ldots, T\lambda_s$ are thermodynamic forces.

Thus, free energy is calculated through the quantity Γ. The above equations allow one to reduce the calculation all mean thermodynamic properties of a complex system to calculation of function Γ. This reduction unifies the problem, but, nevertheless, calculation of the statistical integral for particular systems appears to be a difficult problem. Samples of calculations can be found in many publications.

5.5.2 Unconventional functionals

The derivation of distribution function was based on the expression for entropy, which has the role of a generating functional for the derivation of probability distributions. Though this form of entropy is based on some theoretical consideration, finally, the scheme is justified by empirical evidence. The used method obviously cannot be applied to very small systems. It is possible to assume also, that there are some systems with special type of structure: recursively-like structures or systems with spatial correlations of distant action, or the specific long-term (frozen) correlations, to which classical concepts are inapplicable. It is possible to assume, that in these cases the expression for entropy has another expression, and it is possible to allocate certain areas of inapplicability of the in many respects extremely successful statistical interpretation of the physical reality which is based on the functional (5.26).

Thus, there is an indispensability in variation of the expression for entropy that always should be interpreted as true physical entropy of the thermodynamic system. The first and, probably most known generalization of functional for entropy was offered by Tsallis (1988)

$$S = -\int \rho(\boldsymbol{q}, \boldsymbol{p}) \frac{\rho(\boldsymbol{q}, \boldsymbol{p})^{q-1} - 1}{q - 1} \mathrm{d}\boldsymbol{q} \mathrm{d}\boldsymbol{p}. \tag{5.37}$$

The quantity q can accept positive values. The limiting of the expression for entropy at $q \to 1$ coincides with equation (5.26).

The functional for entropy (5.37) with necessary restrictions can be used for derivation of the function of distribution and calculation of the thermodynamic properties of the physical system.

Apart from the specified above functional (5.37), there were proposed and investigated various functionals for entropy (Luzzi *et al* 2001) that lead to possible statistical interpretation of the thermodynamic system. As it is generating functionals of probability distributions, the procedure initiates particular forms of statistics. It is important to ensure that thermodynamic parities remain unbroken. It is not quite clear whether entropy should always be an additive function.

5.6 Concluding remarks

Let us remark that the most general case that can be investigated by a method of statistical physics is a case of the thermodynamic system in a stationary non-equilibrium state, when internal complexity variables are present and have constant values. There is a problem of extension of the method for consideration of non-stationary processes and for open systems. In the case of open systems, when the number of particles in a system is not constant, the function of Hamilton should be designed specially to take into account the condition of a constancy of the average number of particles.

Thermodynamics and statistical physics are undivisible, because these disciplines study one and the same object (though with different methods)—the systems consisting of many particles—thermodynamic systems. The statistical physics in some cases (not in all) gives molecular interpretation of thermodynamic quantities, but, however, it is impossible to tell, as is sometimes stated, that the statistical physics is the base of thermodynamics. Thermodynamics is formulated independently, and, in fact, the basic principles of thermodynamics appear to be primary in comparison with the heuristic principles of statistical physics.

References

Gibbs J W 1902 *Elementary Principles in Statistical Mechanics: Development with Special Reference to the Rational Foundations of Thermodynamics* (New Haven, CT: Yale University Press)

Jaynes E T 1957 Information theory and statistical mechanics *Phys. Rev.* **106** 620–30

Jaynes E T 1957 Information theory and statistical mechanics *Phys. Rev.* **108** 171–90

Landau L D and Lifshitz E M 1969 *Mechanics* 2nd edn (Oxford: Pergamon)

Landau L D and Lifshitz E M 1986 *Statistical Physics* vol 1 3rd edn (Oxford: Pergamon)

Luzzi R, Vasconcellos A R and Ramos J G 2001 A nonequilibrium statistical ensemble formalism, Maxent-NESOM: basic concepts, construction, application, open questions and criticism *Int. J. Mod. Phys.* B **14** 3089–264

Sommerfeld A 1952 *Thermodynamik und Statistik—Vorlesungen uber theoretische Physik Band 5* (Wiesbaden: Diederichsche Verlagsbuchhandlung) (translation)

Sommerfeld A 1964 *Thermodynamics and Statistical Mechanics—Lectures on Theoretical Physics* vol. V ed F Bopp and J Meixner (New York: Academic) (translated by J Kestin)

Tsallis C 1988 Possible generalization of Boltzmann–Gibbs entropy *J. Stat. Phys.* **52** 479

Part II

From molecules to the living organism

IOP Publishing

Thermodynamics of Complex Systems
Principles and applications
Vladimir N Pokrovskii

Chapter 6

Chemical transformations

Everything that surrounds us, appears to be, as a rule, mixtures of various chemical substances. Many substances can coexist without interaction. Air, for example, is a stable mixture of various substances in gaseous form: oxygen, nitrogen, and other components. But some substances cannot coexist: when such substances are mixed, a chemical reaction begins, resulting in new substances. Humans have learned to purposefully use chemical processes to get the desired effects, for example, by melting metals from ores. There are millions of different chemicals in nature; millions of substances are created artificially. The chemical transformations play a very important role in the process of self-organization of matter, they are important in understanding the problems of the origin and present-day functioning of living objects.

6.1 Kinetics of chemical transformations

6.1.1 Description of reacting mixtures

An example of an unstable mixture is a mixture of hydrogen H_2 and iodine I_2, which are converted to hydrogen iodide. The transformation reaction can be represented by a symbolic equation

$$H_2 + I_2 \rightarrow 2HI$$

which means that a molecule of hydrogen and a molecule of iodine joint together to construct two molecules of iodine hydrogen, so that the composition of the mixture is changing during the process of chemical reaction according to the above rule. The arrow indicates that the chemical reaction is irreversible. By a similar way one can describe the transformation of hydrogen and oxygen into water

$$2H_2 + O_2 \rightarrow 2H_2O$$

doi:10.1088/978-0-7503-3451-8ch6

Generally, the system, which should be considered, is a mixture of k substances $A_1, A_2, \dots, A_k$ with arbitrary numbers of molecules (numbers of moles)[1]

$$n_1,\ n_2,\ \dots,\ n_k. \tag{6.1}$$

Some of the reacting substances can come into the system (into volume V) from the environment or leave the system.

Our task is to describe possible transformations within the system, introducing, first of all, some characteristics of the transformation. For any reaction, as a consequence of balance of substances, a formula can be written—the so-called *stoichiometric formula* in the form

$$\nu_1 A_1 + \nu_2 A_2 + \cdots + \nu_n A_n = 0 \tag{6.2}$$

where A_α is a symbol of the substance α and ν_α is the number of molecules of the substance α participating in the reaction. One has to distinguish the reacting substances and product of reaction. It is convenient to consider that negative values of ν_α are ascribed to reacting substances, whereas the positive values correspond to the products of reaction. For the description of complex reactions running at the same time, one records a few stoichiometric parities.

At a chemical transformation, there is a decrease in numbers (and, accordingly, concentration) of molecules of some types and an increase in numbers (and, accordingly, concentration) of molecules of other types, and the progress of chemical reactions in time can be described by the system of differential equations

$$\frac{dn_\alpha}{dt} = f_\alpha(T, p, n_1, n_2, \dots, n_k), \quad \alpha = 1, 2, \dots, k. \tag{6.3}$$

The set of the reacting substances and the form of functions on the right side of the equations are defined for each chemical reaction empirically. Note that temperature T and pressure p are controlled usually during chemical transformations.

The stoichiometric formula (6.2) is valid both for equilibrium, and for non-equilibrium situations, so that the values of reacting substances in the transformation (6.3) at any time must satisfy equation (6.2). The final state of the system of reacting substances depends on, whether the system cooperates with the environment or remains isolated. In the latter case the reaction comes to the end, and the system comes to chemical equilibrium with values of numbers of molecules

$$n_1^{(o)},\ n_2^{(o)},\ \dots,\ n_k^{(o)}.$$

---

[1] Alternatively, appropriate weight concentration of substances can be set

$$c_\alpha = \frac{m_\alpha}{V} n_\alpha,\ \alpha = 1, 2, \dots, k$$

where m_α—weight of a molecule α; n_α— is the number of molecules of substance α in volume V. The index of substance is designated by the Greek symbol.

The deviation of values of concentration of substances from their equilibrium values can appear to be evidence of non-equilibrium of the system, and it is possible to introduce the quantities describing a deviation of the system from equilibrium

$$\Delta n_\alpha = n_\alpha - n_\alpha^{(o)}, \quad \alpha = 1, 2, \dots, k. \tag{6.4}$$

In virtue of relation (6.2), there are parities between variations of numbers of molecules during the process of reaction: any changes of the number of molecules (it is convenient to consider n_α to be the number of particles or moles of the substance α) should satisfy conditions

$$\frac{\Delta n_1}{\nu_1} = \frac{\Delta n_2}{\nu_2} = \dots = \frac{\Delta n_k}{\nu_k} = \Delta \xi. \tag{6.5}$$

For the characteristic of the running of chemical reaction, one can use only one quantity ξ that, in this case, is considered as the internal variable of the system of reacting substances. The relation (6.5) introduces *extent* ξ or *a degree of incompleteness of chemical reaction*[2]. Due to the choice of signs of the quantities ν_α in equation (6.2), the extent is defined as a positive quantity, whereas $\xi = 0$ in equilibrium.

Let us notice, that now, instead of the relations (6.3), the kinetic equation for the chemical reaction, can be recorded in the form

$$\frac{d\xi}{dt} = f(T, p, \xi).$$

In the case of many reactions, each of them is characterized by a measure of incompleteness ξ_i, and a set of kinetic equations for the coupled reactions can be recorded as

$$\frac{d\xi_i}{dt} = f_i(T, p, \xi_1, \xi_2, \dots, \xi_r), \quad i = 1, 2, \dots, r, \tag{6.6}$$

where i is a label of reaction in the system. The number of internal variables is equal to the number of reactions in the system, that is, certainly, different from the number of reacting substances.

Note that chemical reactions are accompanied by processes of heat generation and movement of substances within the reactor, which leads to the appearance of other internal variables, including temperature and concentration gradients, which should be considered for a more complete description of the non-equilibrium state of the system, in addition to the parameters of completion of chemical reactions.

6.1.2 Examples of elementary reactions

For investigating chemical transformations, the function on the right-side hand of equation (6.3) ought to be specified for each chemical reaction, whereas some types of reactions can be defined (see, for example, Emanuel and Knorre 1969).

---

[2] This quantity was introduced by de Donder (Donder de and van Rysselberghe 1936).

6.1.2.1 Unimolecular reactions

The process of decomposition of a certain substance A or process of transformation of a molecule in molecules of other substances can be described as

$$\mathsf{A} \rightarrow \mathsf{B} + \mathsf{C} + \mathsf{D} + \cdots$$

The rate of decomposition is assumed to be independent on the amount of the resultant substances, so that decomposition of the substance A can be described as

$$\frac{\mathrm{d}n_A}{\mathrm{d}t} = -k_A(T, p)\, n_A, \tag{6.7}$$

and the concentration of the substance changes according the law

$$n_A(t) = n_A(0)\exp[-k_A(T, p)\, t].$$

The reactions of this type are classified as chemical reactions of the first order.

6.1.2.2 Bimolecular reactions

When a reaction of composition of a new substance from two different substances is considered, according to the scheme

$$\mathsf{A} + \mathsf{B} \rightarrow \mathsf{C},$$

one can imagine that possibility of emerging of molecule C is connected with the number of collisions of molecule A with molecule B, which is the bigger the bigger the concentrations of the substances, so that one can write for the change of numbers of the substances

$$\begin{aligned} \frac{\mathrm{d}n_A}{\mathrm{d}t} &= -\, k_A(T, p)\, n_A n_B, \\ \frac{\mathrm{d}n_B}{\mathrm{d}t} &= -\, k_B(T, p)\, n_A n_B, \\ \frac{\mathrm{d}n_C}{\mathrm{d}t} &= k_C(T, p)\, n_A n_B. \end{aligned} \tag{6.8}$$

In virtue of stoichiometric relation, all coefficients are equal to each other

$$k_A = k_B = k_C = k(T, p).$$

The reactions of this type, when the right-hand side of the kinetic equation contains concentrations in mutual second order, are classified as chemical reactions of the second order.

6.1.2.3 Trimolecular reactions

When three molecules are participating in a reaction of composition of a new substance, for example, according to the scheme

$$2\mathsf{A} + \mathsf{B} \rightarrow \mathsf{C},$$

one can imagine that possibility of emerging of molecule C is connected with the number of collisions of two molecules of A with molecule B, so that, according to the previous speculations, one can write for the change of the numbers of the substances

$$\begin{aligned}\frac{dn_A}{dt} &= -2k(T, p)\, n_A^2 n_B,\\ \frac{dn_B}{dt} &= -k(T, p)\, n_A^2 n_B,\\ \frac{dn_C}{dt} &= k(T, p)\, n_A^2 n_B.\end{aligned} \tag{6.9}$$

The stoichiometric relation is taken into account already.

The reactions of this type, when the right-hand side of the kinetic equation contains concentrations in mutual third order, are classified as chemical reactions of the third order.

6.1.2.4 Complex reactions

In a similar way one could define reactions of higher orders, for example, reactions of the fourth order, according to the scheme

$$2\mathrm{A} + \mathrm{C} + \mathrm{B} \rightarrow \mathrm{D}.$$

However, probability of simultaneous collision of four and more molecules is small, and the mechanism of such a reaction could be different.

The substances can react in a few stages, for example, substance A transforms to substance C, but with an intermediate substance B,

$$\mathrm{A} \rightarrow \mathrm{B}, \quad \mathrm{B} \rightarrow \mathrm{C},$$

so that one can write, following the rules for elementary reactions, for the change of the numbers of the substances

$$\begin{aligned}\frac{dn_A}{dt} &= -k_A(T, p)\, n_A,\\ \frac{dn_B}{dt} &= k_A(T, p)\, n_A - k_B(T, p)\, n_B,\\ \frac{dn_C}{dt} &= k_B(T, p)\, n_B.\end{aligned} \tag{6.10}$$

It is easy to see that the rate of transformation of substance A into substance C depends on both the coefficient k_A and the coefficient k_B. In a simple case, when concentration of substance A is maintaining constant, $c_A = c_A^0$, concentration of the intermediate substance is

$$n_B = \frac{k_A}{k_B} n_A^0$$

and the concentration of substance C increases linearly. In the case, when the kinetic coefficients are equal to each other

$$k_A = k_B,$$

intermediate products can exist.

This simple example shows us difficulties, arising, when mechanisms of chemical reactions are investigated. The situation appears to be rather complicated, so that each chemical transformation requires its own careful examination. The elucidation of mechanism of a complex reaction is a difficult and tricky enterprise and may lead to diverging opinions among investigators.

6.2 Thermodynamics of chemical transformations

We consider a reactor of a certain volume V, in which molecules of different substances are present and are participating in r independent or coupled chemical reactions, whereas any substance can participate in any reaction. We introduce symbol $n_\alpha^{(j)}$ for the number of molecules (moles) of the kind α, participating in a reaction labeled j, whereas we have for the total number of molecules (moles) n_α in the system the expression

$$n_\alpha = \sum_{j=1}^{r} n_\alpha^{(j)}.$$

In any case one can use a set of variables n_α ($\alpha = 1, 2, \ldots, k$) to describe the composition, but, if one considers a system with internal chemical reactions, one ought to separate the contribution of chemical reactions in the change of the numbers of particles. The changes of the variables occur due to chemical transformations and fluxes of substances into and out of the system through the borders of the volume. In virtue of relations (6.5) (for each chemical reaction $j = 1, 2, \ldots, r$), not all of the numbers n_α appear to be independent variables, and one has to write

$$\Delta n_\alpha = \sum_{j=1}^{r} \nu_\alpha^{(j)} \Delta\xi_j + \Delta^e n_\alpha, \tag{6.11}$$

where $\Delta^e n_\alpha$ is a number of particles of the kind α having come through the borders into the system. Some substances, participating in chemical reactions, are coming into the system outside ($\Delta^e n_\alpha > 0$) and some substances leave the volume of the reactor ($\Delta^e n_\alpha < 0$).

Relations (6.11) contain two possible sets of variables for description of thermodynamic characteristics; the independent variables of state appear to be the degrees of incompleteness of chemical reactions ξ_i ($i = 1, 2, \ldots, r$), which in this situation are considered as internal variables, and the numbers of participating molecules that are coming through the borders of the reactor into the system n_α. Due to the stoichiometric relations, the intermediate substances are excluded from the set of independent variables.

For the creation of a thermodynamic description of systems with chemical reactions we are largely obliged to the Brussels school of de Donde (Donder de and van Rysselberghe 1936, Prigogine and Defay 1954, Prigogine 1961). Thanks to these researchers, we can formulate the thermodynamics of chemical transformations.

6.2.1 Thermodynamic potential and affinity

It is assumed that the system is controlled by the temperature T and pressure p, so that it is convenient for the system to be characterized by the Gibbs free energy or thermodynamic potential $G = F + pV$, where F is free energy, defined in section 2.4.2. The total differential appears to be a particular case of equation (2.20), and, in the situations when constant temperature and pressure are considered, the Gibbs free energy (thermodynamic potential) of the system can be represented as a function of composition, that is,

$$dG = \sum_{\alpha=1}^{k} \mu_\alpha \, \Delta n_\alpha,$$

The quantities Δn_α in this relation present total changes of the numbers of molecules of substance labeled α. Due to internal chemical reactions, not all of these quantities appear to be independent variables. We assume that the deviation of the system from equilibrium is completely characterized with the possible incompleteness of the chemical reaction, and the contribution of other internal variables, described by the middle term in equation (2.20), is neglected.

Using the quantities Δn_α, defined by equation (6.11), one can obtain for any situation at the constant temperature and pressure

$$dG = -\sum_{j-1}^{r} A_j \, \Delta\xi_j + \sum_{\alpha=1}^{k} \mu_\alpha \, \Delta^e n_\alpha, \qquad A_j = -\sum_{\alpha=1}^{k} \mu_\alpha \nu_\alpha^{(j)}. \tag{6.12}$$

Thus, in this situation, the incompletenesses of chemical reactions ξ_i $(i = 1, 2, \ldots, r)$ and numbers of molecules that are coming from the environment, appears to be independent thermodynamic variables.

The quantity A_j in equations (6.12) is called *affinity*, as it was dubbed by de Donde. The chemical potentials of substances μ_α can be expanded in series with regard to concentrations of different substances in the vicinity of equilibrium point, so that one eventually can get the dependence of the affinities on the internal variables

$$A_i = -\sum_{j=1}^{r} \sum_{\alpha,\gamma=1}^{k} \frac{\partial \mu_\alpha}{\partial c_\gamma} \nu_\alpha^{(i)} \nu_\gamma^{(j)} \Delta\xi_j + \cdots, \qquad j = 1, 2, \ldots, r.$$

The affinities in equilibrium are equal to zero, $A_j = 0$.

The closed system, when there are no fluxes of substances through the borders, reaches the equilibrium situation, and all thermodynamic functions are constant. In the case of the open system, when there are fluxes of substances into and out through the borders, the system cannot reach equilibrium state, but, in an unchanged environment, the system can stay in a stationary state, when there are also no changes of thermodynamic functions. The condition of such situations, as it follows equation (6.12), is the balance between the change of energy of the system (due to the reactions) and incoming flux of energy, carried by fluxes of substances

$$\sum_{j=1}^{r} A_j \, \Delta\xi_j = \sum_{\alpha=1}^{k} \mu_\alpha \, \Delta^e n_\alpha. \tag{6.13}$$

There are fluxes of matter through the system, but the composition of the system appears unchanged. A non-equilibrium state of the system is fixed with internal variables ξ_j, that are incompletenesses of chemical reactions. Note that it is assumed that all times of relaxations of internal variables are less than the time of the process.

6.2.2 Entropy of the reacting mixture

Considering a particular case of the general parity (2.8), the change of entropy S of the system, in which only deviations from equilibrium connected with chemical transformations, can be recorded as

$$T\,dS = \Delta Q + \sum_{\alpha=1}^{k} \eta_\alpha \, \Delta n_\alpha. \tag{6.14}$$

The chemical reactions are accompanied by emission ($\Delta Q < 0$) or absorption ($\Delta Q > 0$) of heat.

Via equations (6.11), we introduce the independent variables: incompleteness of chemical reaction ξ_i ($i = 1, 2, \ldots, r$) and number of particles n_α ($\alpha = 1, 2, \ldots, k$), coming into the system from the environment, to obtain

$$T\,dS = \Delta Q + \sum_{j=1}^{r} \sum_{\alpha=1}^{k} \eta_\alpha \nu_\alpha^{(j)} \, \Delta\xi_j + \sum_{\alpha=1}^{k} \eta_\alpha \, \Delta^e n_\alpha. \tag{6.15}$$

Comparing equation (6.15) with general definition of total alternation of entropy (2.8) allows us to define the thermodynamic force, associated with internal variable ξ_j

$$\Xi_j = -\sum_{\alpha=1}^{k} \eta_\alpha \nu_\alpha^{(j)}, \quad j = 1, 2, \ldots, r. \tag{6.16}$$

The middle term of equation (6.15) presents dissipation of energy due to relaxation of internal variables that are incompletenesses of chemical reactions.

In a stationary state, entropy, similar to other thermodynamic functions, does not change, so equation (6.15) determines a heat, received by the system due to the reactions and external fluxes

$$\Delta Q = \sum_{j=1}^{r} \Xi_j \, \Delta\xi_j - \sum_{\alpha=1}^{k} \eta_\alpha \, \Delta^e n_\alpha.$$

Due to relation (2.27), $\eta_\alpha = h_\alpha - \mu_\alpha$, so that the constancy of thermodynamic potential in stationary states allows us to change the quantity η_α in the above equation for enthalpy h_α, and to get the equation of thermal balance of the system for the stationary case

$$\Delta Q - \sum_{j=1}^{r} (A_j + \Xi_j) \, \Delta\xi_j + \sum_{\alpha=1}^{k} h_\alpha \, \Delta^e n_\alpha = 0. \tag{6.17}$$

The chemical reactions are accompanied by processes of emerging of heat and motion of substances within the reactor, and, for a more complete description of the situation, one needs to consider, in addition to completeness of chemical reactions, other internal variables, which are, for example, gradients of temperature and concentration.

6.2.3 Example of non-equilibrium steady state

Further in this section, as a simplest example, we shall consider, following the author's work (Pokrovskii 2005), a reactor, in which there is a reaction of association–dissociation that can be represented by symbolic equation

$$2\,\mathsf{A} \leftrightarrow \mathsf{B}.$$

It means that two molecules of substance A (mass m) can unite in one molecule of substance B (mass $2m$), whereas a molecule of substance B can split in two molecules of substance A, so that changes of concentrations of substances are connected with each other

$$2\, dn_A = -dn_B, \qquad dc_A = -dc_B. \tag{6.18}$$

Each substance in the system is characterized by a chemical potential and specific enthalpy; the equilibrium condition, due to the above relations, can be represented as

$$\eta_A(c_A, c_B) = \eta_B(c_A, c_B).$$

This equation defines the equilibrium concentrations of substances c_A^0 and c_B^0, if, for example, the total mass of substances is given. To consider approaching the equilibrium state, we need to refer to the equations of chemical kinetics. We can imagine that there is the incoming flux J of substance A that transforms in substance

B during the reaction, while substance B permanently goes out of the reactor, so that equations for concentrations of substances can be written as

$$\frac{dc_A}{dt} = J - \kappa_{[A\to B]}c_A^2 + \kappa_{[B\to A]}c_B,$$
$$\frac{dc_B}{dt} = \kappa_{[A\to B]}c_A^2 - \kappa_{[B\to A]}c_B - J. \tag{6.19}$$

Under the equilibrium situation ($J = 0$), the time derivatives are equal to zero, and one can find the relation between the kinetic constants $\kappa_{[A\to B]}$ and $\kappa_{[B\to A]}$ and the equilibrium concentrations

$$\kappa_{[B\to A]}c_B^\circ = \kappa_{[A\to B]}(c_A^\circ)^2.$$

The differences between non-equilibrium and equilibrium concentrations $\Delta c_A = c_A - c_A^\circ$ and $\Delta c_B = c_B - c_B^\circ$ can be chosen to describe deviation of the state of the system from equilibrium, while $\Delta c_A = -\Delta c_B$. In this case, equations (6.19) are followed

$$\frac{dc_A}{dt} = J - \frac{1}{\tau}\Delta c_A, \quad \frac{1}{\tau} = \kappa_{[A\to B]}c_A^\circ\left(2 + \frac{c_A^\circ}{c_B^\circ}\right). \tag{6.20}$$

In a steady-state situation, the left-hand sides of the kinetic equation are equal to zero, so that one can determine a deviation of concentration from equilibrium

$$\Delta c_A = \tau J. \tag{6.21}$$

In virtue of equation (6.14), the change of entropy in unit of time due to the flux of substances is determined as

$$\frac{\mathrm{d}_e S}{\mathrm{d}t} = \frac{1}{T}(\eta_A - \eta_B)J.$$

It is convenient to use an expansion of specific quantities $\eta_A(c_A, c_B)$ and $\eta_B(c_A, c_B)$ and, taking into account that in equilibrium $\eta_A(c_A^\circ, c_B^\circ) = \eta_B(c_A^\circ, c_B^\circ)$, write the difference as

$$\eta_A - \eta_B = -B\,\Delta c_A, \quad B = \left(\frac{\partial \eta_A}{\partial c_A} - \frac{\partial \eta_A}{\partial c_B} - \frac{\partial \eta_B}{\partial c_A} + \frac{\partial \mu_B}{\partial c_B}\right).$$

In steady-state situation, the external flux of entropy is compensated by the internal production of entropy

$$\frac{\mathrm{d}_i S}{\mathrm{d}t} = -\frac{\mathrm{d}_e S}{\mathrm{d}t} = \frac{1}{T}\frac{B}{\tau}(\Delta c_A)^2. \tag{6.22}$$

One can see that there is permanent internal production of entropy which is connected with deviation of concentration from equilibrium Δc_B. The further speculations are quite similar to those demonstrated in section 4.6 of chapter 4, so that one can define the thermodynamic force

$$\Xi = \frac{1}{T}\frac{B}{\tau}\,\Delta c_A, \qquad \xi = \Delta c_A. \tag{6.23}$$

and, at the end, calculate entropy of the system in the non-equilibrium state

$$S - S_0 = -\frac{B}{2\tau T}\,(\Delta c_A)^2. \tag{6.24}$$

One can use equation (6.21) for the flux of substance in and out of the system to rewrite the expression for entropy in the form

$$S - S_0 = -\frac{\tau B}{2T}\,J^2.$$

6.3 Brusselator

While investigating chemical transformation, a phenomenon of space–time periodicity was discovered. This phenomenon can be illustrated on the basis of a simple model, proposed and investigated by Prigogine with collaborators in 1968 (Nicolis and Prigogine 1977).

The transformations of the considered substances can be represented symbolically in the form

$$\begin{aligned} {}^{\downarrow}\mathsf{A} &\to \mathsf{X} \\ {}^{\downarrow}\mathsf{B} + \mathsf{X} &\to \mathsf{Y} + \mathsf{D}_{\downarrow} \\ 2\mathsf{X} + \mathsf{Y} &\to 3\mathsf{X} \\ \mathsf{X} &\to \mathsf{E}_{\downarrow} \end{aligned} \tag{6.25}$$

The substances A and B are coming into a reactor. Through intermediate substances X and Y, the incoming substances are transforming to substances D and E, which are being removed from the reactor.

The set of reactions (6.25) defines the object that is called brusselator, the name invented on the sample of the name of a mathematical pendulum—oscillator. The special attention to the space–time periodicity was caught with a connection with the hope of coming close to understanding functioning of biological organisms. The dramatic history of an early stage of research of the phenomenon in the Soviet Union was described by a participant of events (Shnol 1997, chapter 10).

6.3.1 A chemical clock

Considering the set of reactions (6.25), we investigate firstly the case when the fluxes of the substances A and B are given and the all substances are distributed in the reactor homogeneously. It could be expected that there is some steady state of system with constant values of intermediate reagents. How does concentration of substances X and Y change at a deviation from a stationary point?

According to the set of chemical transformations (6.25), the system of the equations for concentration of intermediate substances in the reactor should be recorded in the form

$$\frac{d[X]}{dt} = k_1[A] - k_2[B][X] + k_3[X]^2[Y] - k_4[X],$$
$$\frac{d[Y]}{dt} = k_2[B][X] - k_3[X]^2[Y]. \tag{6.26}$$

The symbol of a substance in square brackets depicts concentration of the substance, that is the number of molecules (moles) per unit of volume; the symbols k_1, k_2, k_3, k_4 are constants of chemical reactions.

Further on, it is more convenient to use the new time $t = k_4 t_{old}$ and new variables

$$A = \left(\frac{k_1^2 k_3}{k_4}\right)^{\frac{1}{2}}[A], \quad B = \frac{k_2}{k_4}[B],$$
$$X = \left(\frac{k_3}{k_4}\right)^{\frac{1}{2}}[X], \quad Y = \left(\frac{k_3}{k_4}\right)^{\frac{1}{2}}[Y]. \tag{6.27}$$

The set of equations (6.26) in the new variables takes a more simple form

$$\frac{dX}{dt} = A - BX + X^2Y - X,$$
$$\frac{dY}{dt} = BX - X^2Y. \tag{6.28}$$

The quantities A and B appear to be the parameters of the problem. As the initial conditions, arbitrary values of variables X and Y can be given.

It is easy to see that a steady-state point of the system is

$$X = A, \quad Y = \frac{B}{A}. \tag{6.29}$$

To investigate behaviour of the system near the steady-state point, we consider time dependent solutions of equations (6.28), using the representation

$$X(t) = A(1 + x(t)), \quad x(t) \ll 1,$$
$$Y(t) = \frac{B}{A}(1 + y(t)), \quad y(t) \ll 1. \tag{6.30}$$

With proximity up to the first order, from equations (6.28), one has the equations for the small quantities

$$\frac{dx}{dt} = (B - 1)x + By,$$
$$\frac{dy}{dt} = -A^2x - A^2y. \tag{6.31}$$

Let us notice, that the variables x and y can be interpreted as internal variables of the system, and it is possible to compare equations (6.31) with the general equations for dynamics of internal variables (3.4). Here we encounter the case, when the matrix R_{ij} in the equation (3.4) (at small values of internal variables) is not positively determined at all values of parameters.

One can assume the solution of equations (6.31) in the form

$$x(t) = x(0)\, e^{\lambda t}, \quad y(t) = y(0)\, e^{\lambda t},$$

where $x(0)$ and $y(0)$ are initial values of variables, which are obeyed by the system of algebraic equations

$$(B - 1 - \lambda)\, x(0) + B\, y(0) = 0,$$
$$-A^2\, x(0) - (A^2 + \lambda)\, y(0) = 0. \tag{6.32}$$

The condition of existence of a non-trivial solution

$$\lambda^2 + (1 + A^2 - B)\lambda + A^2 = 0$$

defines the value of the index

$$\lambda = -\frac{1 + A^2 - B}{2} \pm \sqrt{\frac{(1 + A^2 - B)^2}{4} - A^2}\,.$$

It is possible to notice, that approaching of the system to the stationary point (6.29), that is, the changes of concentration of intermediate substances, depends on values of parameters of system A and B. In the case when $B = 1 + A^2$, there are pure oscillations according to the law

$$x(t) = x(0)\, e^{\pm iAt}, \quad y(t) = y(0)\, e^{\pm iAt}.$$

In the considered situation, one can observe pulsations near the steady-state point (6.29), and the set of equations (6.25) can be considered as an elementary chemical clock.

6.3.2 Kinetics of brusselator

Now, coming back to the description of chemical transformations (6.25), we assume, that substances A and B appear to spread homogeneously inside the reactor, while concentrations of the intermediate substances X and Y can change in time and in space according to the chemical transformations and law of diffusion.

For the sake of simplicity, one can consider only the processes in one dimension, so that a set of equations of chemical transformations and diffusion in the reactor can be written as

$$\frac{\partial[\mathrm{X}]}{\partial t} = k_1[\mathrm{A}] - k_2[\mathrm{B}][\mathrm{X}] + k_3[\mathrm{X}]^2[\mathrm{Y}] - k_4[\mathrm{X}] + \mathcal{D}_X\frac{\partial^2[\mathrm{X}]}{\partial z^2},$$
$$\frac{\partial[\mathrm{Y}]}{\partial t} = k_2[\mathrm{B}][\mathrm{X}] - k_3[\mathrm{X}]^2[\mathrm{Y}] + \mathcal{D}_Y\frac{\partial^2[\mathrm{Y}]}{\partial z^2}. \tag{6.33}$$

Similar to the previous section, the symbol of a substance in square brackets depicts concentration of the substance, that is the number of molecules (moles) per unit of volume; the symbols k_1, k_2, k_3, k_4 are constants of chemical reactions and $\mathcal{D}_\mathrm{X}$ and $\mathcal{D}_\mathrm{Y}$ are coefficients of diffusion of the intermediate substances X and Y.

Further on, we shall use the new time $t = k_4 t_{\mathrm{old}}$ and the non-dimensional variables, introduced by relations (6.27). It is convenient also to define the new coefficients of diffusion as

$$D_\mathrm{X} = \frac{\mathcal{D}_\mathrm{X}}{k_4}, \qquad D_\mathrm{Y} = \frac{\mathcal{D}_\mathrm{Y}}{k_4},$$

so that the set of equations (6.33) takes the simpler form

$$\frac{\partial X}{\partial t} = A - BX + X^2Y - X + D_\mathrm{X}\frac{\partial^2 X}{\partial z^2},$$
$$\frac{\partial Y}{\partial t} = BX - X^2Y + D_\mathrm{Y}\frac{\partial^2 Y}{\partial z^2}.$$

To simplify the writing of the equations more, one can denote partial derivatives by symbol of variable with low indexes of variable in respect of which one differentiates. So one can rewrite the above set of equations as

$$X_t = A - BX + X^2Y - X + D_\mathrm{X}X_{zz},$$
$$Y_t = BX - X^2Y + D_\mathrm{Y}Y_{zz}. \tag{6.34}$$

It is natural to consider that there are no fluxes of the intermediate substances X and Y through the borders of the reactor. So, one puts on the border conditions

$$X_z(t, 0) = X_z(t, L) = 0,$$
$$Y_z(t, 0) = Y_z(t, L) = 0. \tag{6.35}$$

As the initial conditions, arbitrary dependence of X and Y on the co-ordinate can be given.

It was assumed that the quantities A and B do not depend on co-ordinate and time, so that the brusselator, as a thermodynamic system, can be found in one of its stationary states. In this case, the variables X and Y do not depend on time and can only be functions of co-ordinates, so that the set of equations (6.34) can be rewritten as

$$A - BX + X^2Y - X + D_\mathrm{X}X_{zz} = 0,$$
$$BX - X^2Y + D_\mathrm{Y}Y_{zz} = 0. \tag{6.36}$$

It is easy to see that the simplest steady-state solution is

$$X = A, \quad Y = \frac{B}{A}. \tag{6.37}$$

Alongside this state, there is a set of solutions of equations (6.36), each of which describes the non-uniform steady state of the thermodynamic system. For simplicity, we shall consider solutions of equations (6.34) near the steady-state point (6.37) to demonstrate the existence of non-uniform steady states. To investigate this situation, one can use the presentation of a solution of equations (6.34)

$$\begin{aligned} X(t, z) &= A + x(t, z), \quad x(t, z) \ll A, \\ Y(t, z) &= \frac{B}{A} + y(t, z), \quad y(t, z) \ll \frac{B}{A}. \end{aligned} \tag{6.38}$$

With proximity up to the first order, one has from equations (6.34), the equations for the small quantities

$$\begin{aligned} x_t &= (B - 1)x + A^2 y + D_X x_{zz}, \\ y_t &= -Bx - A^2 y + D_Y y_{zz} \end{aligned} \tag{6.39}$$

with the border and initial conditions

$$\begin{aligned} x_z(t, 0) = x_z(t, L) = 0, \quad x(0, z) = \phi(z), \\ y_z(t, 0) = y_z(t, L) = 0, \quad y(0, z) = \psi(z). \end{aligned} \tag{6.40}$$

The arbitrary functions $\phi(z)$ and $\psi(z)$ are determined by perturbation of an initial state of thermodynamic system, and one can consider these functions to describe thermal fluctuation of concentrations.

It is easy to be convinced that the set of linear equations (6.39) with border conditions (6.40) has a set of particular solutions of the kind

$$\begin{aligned} x_m(t, z) &= p_m \, e^{\lambda_m t} \cos \frac{m\pi z}{L}, \\ y_m(t, z) &= q_m \, e^{\lambda_m t} \cos \frac{m\pi z}{L}, \end{aligned} \tag{6.41}$$

where p_m and q_m are amplitude of modes, m is an integer number.

A general solution can be written as a sum of the particular solutions

$$\begin{aligned} x(t, z) &= \sum_{m=0}^{\infty} p_m \, e^{\lambda_m t} \cos \frac{m\pi z}{L}, \\ y(t, z) &= \sum_{m=0}^{\infty} q_m \, e^{\lambda_m t} \cos \frac{m\pi z}{L}. \end{aligned} \tag{6.42}$$

These equations describe a set of non-uniform oscillating deviations from the uniform solution (6.37) as a sum of separate harmonics. The amplitudes p_m and q_m must satisfy the initial conditions

$$\phi(z) = \sum_{m=0}^{\infty} p_m \cos \frac{m\pi z}{L},$$
$$\psi(z) = \sum_{m=0}^{\infty} q_m \cos \frac{m\pi z}{L}.$$

Among solutions of equations (6.39), one can separate non-uniform stationary states, the existence of which depends on behaviour of the system near the stationary point (6.37) at preset values of quantities A and B. To investigate the situation, we refer to equation (6.39) and solution (6.41) to formulate a set of algebraic equations for unknown amplitudes p_m and q_m as

$$\left[\lambda_m + 1 - B + D_X\left(\frac{m\pi}{L}\right)^2\right]p_m - A^2 q_m = 0$$
$$B\, p_m + \left[\lambda_m + A^2 + D_Y\left(\frac{m\pi}{L}\right)^2\right]q_m = 0.$$

The system has a non-trivial solution in the case, if the determinant of the system is equal to zero. This requirement determines an equation for quantity λ_m

$$\lambda_m^2 + b_m\lambda_m + c_m = 0,$$
$$b_m = 1 + A^2 - B + (D_X + D_Y)\left(\frac{m\pi}{L}\right)^2,$$
$$c_m = \left[1 + D_X\left(\frac{m\pi}{L}\right)^2\right]\left[A^2 + D_Y\left(\frac{m\pi}{L}\right)^2\right] - B\, D_Y\left(\frac{m\pi}{L}\right)^2.$$

The equation has a simple solution

$$\lambda_m = -\frac{b_m}{2} \pm \sqrt{\frac{b_m^2}{4} - c_m}. \tag{6.43}$$

It is easy to see that value of the real part $\boldsymbol{R}\mathrm{e}\lambda_m$ of the quantity λ_m is negative at small values of the quantity B, and is positive at greater values of B, so that there is a certain critical value $B_c(m)$ for each mode m. The steady-state uniform solution (6.37) is stable at $B < B_c$ and is unstable at $B > B_c$, and in the latter case the system can pass to other stationary stable solutions of equations (6.34), which could appear to be spatially non-uniform. Which of the steady-state solutions is realized depends on initial conditions (6.42) at given values of quantities A and B. The investigation of this situation can be found in the literature (Ebeling 1976, Nicolis and Prigogine 1977).

6.3.3 Thermodynamics of brusselator

The brusselator is considered as a reactor with the set of four chemical reactions (6.25), in which molecules of six substances participate, and now we shall try to describe the behavior of the brusselator under changes of external conditions from the thermodynamic point of view, developed in section 6.2. We assume that the reactions are numerated by index $j = 1, 2, 3, 4$ and the substances—by index $\alpha = \mathsf{A, B, X, Y, D, E}$ and start with equations for thermodynamic potential and entropy of the system (6.12) and (6.15), which, in application to the considered case, gets a form

$$dG = -\sum_{j=1}^{4} A_j \, \Delta\xi_j + \sum_{\alpha=1}^{6} \mu_\alpha \, \Delta^e n_\alpha, \; A_j = -\sum_{\alpha=1}^{6} \mu_\alpha \, \nu_\alpha^{(j)}, \tag{6.44}$$

$$T \, dS = \Delta Q - \sum_{j=1}^{4} \Xi_j \, \Delta\xi_j + \sum_{\alpha=1}^{6} \eta_\alpha \, \Delta^e n_\alpha, \; \Xi_j = -\sum_{\alpha=1}^{6} \eta_\alpha \nu_\alpha^{(j)}. \tag{6.45}$$

The current thermodynamic state of the brusselator is determined by the independent variables: degrees of completeness of chemical reactions ξ_1, ξ_2, ξ_3, ξ_4 and the number of participating molecules, coming through the borders of the reactor, n_A, n_B, n_D, n_E. Due to the stoichiometric relations, the intermediate substances, n_X and n_Y are excluded from the set of independent variables. For simplicity we consider uniform states of the system, similar to that discussed in section 6.3.1. In the general case, we have to consider the listed thermodynamic variables to depend on space variables and to consider equations (6.44) and (6.45) as local relations for density of thermodynamic functions in a certain point of the reactor. In these cases, we have to take into account the diffusion processes.

The fundamental functional structure of the system with chemical reactions is defined completely by stoichiometric numbers $\nu_\alpha^{(j)}$ that, for brusselator can be easily found from equations (6.25). Apart of the stoichiometric numbers $\nu_\alpha^{(j)}$, the expressions (6.44) and (6.45) contain characteristics of the system: the chemical potentials μ_α and the contribution into entropy for unit of the incoming substances η_α / T. These quantities depend on concentrations of the substances and, in the general case, due to the possible non-uniformity of the reactor, on space variable z.

The affinity of each reaction A_i can be presented, according to the known values of $\nu_\alpha^{(j)}$ and the second equation from the set (6.44), as

$$\begin{aligned} A_1 &= \mu_A - \mu_X, \\ A_2 &= \mu_B + \mu_X - \mu_Y - \mu_D, \\ A_3 &= \mu_Y - \mu_X, \\ A_4 &= \mu_X - \mu_E. \end{aligned} \tag{6.46}$$

The chemical potentials μ_A, μ_B, μ_X, μ_Y, μ_D, μ_E of the substances participating in reactions depend on the numbers of molecules of all substances and can be expanded over these quantities near the equilibrium point, in which affinity is equal to zero.

The change of numbers of reacting substances within the reactor is determined by equations (6.11), and, for the brusselator, can be determined directly from (6.25) as the alternations of internal variables (degrees of completeness of reactions) and changes of the particle of reacting substances, coming through the borders of the reactor

$$
\begin{aligned}
\Delta n_A &= -\,\Delta\xi_1 + \Delta^e n_A,\\
\Delta n_B &= -\,\Delta\xi_2 + \Delta^e n_B,\\
\Delta n_X &= \Delta\xi_1 - \Delta\xi_2 + \Delta\xi_3 - \Delta\xi_4,\\
\Delta n_Y &= \Delta\xi_2 - \Delta\xi_3,\\
\Delta n_D &= \Delta\xi_2 - \Delta^e n_D,\\
\Delta n_E &= \Delta\xi_4 - \Delta^e n_E.
\end{aligned}
\tag{6.47}
$$

At the constant fluxes of the incoming and outcoming substances, one can find the brusselator in one of the steady states, among them a unique state with constant values of intermediate reagents (6.29). In stationary situations, relations (6.47) become simpler and allow one to connect the change of internal variables with the change of reacting substances

$$
\begin{aligned}
\Delta\xi_1 &= -\,\Delta n_A,\\
\Delta\xi_2 &= -\,\Delta n_B = \Delta n_D,\\
\Delta\xi_3 &= \Delta n_X + \Delta n_A - \Delta n_B + \Delta n_E = -\Delta n_Y - \Delta n_B,\\
\Delta\xi_4 &= \Delta n_E.
\end{aligned}
\tag{6.48}
$$

These relations do not define that the variables ought to be constant; the pulsating changes appear to be possible.

Consider the change of state of the system from one state to another, when the quantities of introduced substances A and B are increasing or decreasing, while the system does not leave the steady-state situation described by (6.29). We can assume that Gibbs free energy during the transition keeps its minimal value, so that equation (6.44) is followed by equation (6.13), which connects the independent variables. Then, according to equation (6.45), entropy of the system can be calculated, taking into account the above-discussed formulae. We cannot calculate the changes of entropy, so the results depend on numerical values of chemical potentials and enthalpy of the substances. We can assume, that the total change of entropy of the system can be positive or negative; it depends on the relations among the parameters of the system, which are functions of concentrations of all substances. It looks natural, if the variation of entropy is positive. However, the situation could be surprising, if, above the critical point the system could prefer to choose another steady state with lower value of entropy.

Let observe that expressions (6.44) and (6.45) include only internal processes connected with chemical reactions. One can notice, that, when the distribution of substances in space is non-uniform, it is also necessary to take into consideration diffusive processes; in addition to degrees of completeness of chemical reactions, there are local gradients of concentration of substances.

6.4 The problem of self-organization

The described example of the brusselator demonstrates the phenomenon of spontaneous emergence of space structure within a volume with a set of chemical reactions. Under increasing flux of incoming substances, at a certain critical point, the space-uniform state of the brusselator appears to be unstable, and changes to one of the non-uniform steady states, described in section 6.3.3. These stable structures, appearing above the critical point, were called *dissipative structures* (Nicolis and Prigogine 1977). The form of dissipative structure depends on modes (fluctuations) that were initially agitated, while the structure can change its form with changing of the amount of coming substances. According to some estimates (Nicolis and Prigogine 1977, chapter 8), entropy of the brusselator decreases after the transition to the new state, which is, at first glance, surprising.

To describe the phenomenon from the thermodynamic point of view, we shall estimate the change of the characteristic functions during the transition, considering the system to be in steady states before and after the transition. The transition itself is a non-stationary process that is controlled, as was said earlier, by Gibbs free energy (thermodynamic potential) that is a leading control quantity for all transformation in the system of running chemical reactions at temperature T and pressure p being constant. The potential is tending to its minimum value, and it means that the terms in equation (6.12), which presents the contribution into thermodynamic potential from internal variables, is tending (with T and p constant) to disappear, and we believe that the contribution from these terms appears to be less after the transition.

To estimate the change of entropy of the system during the transition, one has to refer to equation (6.15). All terms on the right-hand side of the equation are changing; however, in a stationary state, before and after the transition, there is a balance of process of the growth of entropy of the system due to the incoming stream of substances and process of decrease of entropy due to outgoing stream of heat. Considering the process of transformation, one has to take into account the processes of emerging and relaxation of internal variables. During the process of transition to the non-uniform state a new set of internal variables with very great times of relaxation (the dissipative structures) has been emerging, which brings a decrease of entropy. The relaxation of these internal variables (the dissipative structures) and, consequently, the dissipation of energy (production of entropy) from such internal variables would be delayed (or stopped in the case, if the relaxation time is infinity), and the dissipation of energy of the system will not correspond to incoming flux of energy, and the difference can be defined as the quantity

$$\frac{1}{T}\sum_i \Xi_i \, d\xi_i, \tag{6.49}$$

where the summing is providing over the variables with very great relaxation times. Formally, the quantity (6.49) can be considered as a negative contribution into entropy during the transition. New balance after transition occurs at the new value of entropy that can be lower than the value before the transition.

It is amazing that there is such a mechanism of *delayed relaxation* of the internal variables. Due to it, entropy can decrease during a process of transition to a new state. But it cannot be said that the entropy always decreases. The condition of it is the emerging of internal variables with greater relaxation times, which means the creation of some structure. The decrease of entropy is connected with the increasing of complexity (in the sense of section 8.2.2) of the system; the structure of the system becomes more complex, and one can say about the transition as about a self-organizing process. One can say that the decreasing of entropy and, correspondingly, the increasing of complexity is *a natural process* within the system of reacting substances, when there is an abundance of available energy.

References

Donder de T and van Rysselberghe P 1936 *Thermodynamic Theory of Affinity: A Book of Principles* (Stanford, CA: Stanford University Press)

Ebeling W 1976 *Strukturbildung bei Irreversiblen Prozessen* (Leipzig: BSB B.G. Teubner Verlagsgesellschaft)

Emanuel N M and Knorre D G 1969 *Kurs khimicheskoy kinetiki, Izdanie vtoroe, ispravlennoe i dopolnennoe [The Course of Chemical Kinetics]* 2nd edn corrected and completed (Moscow: Vysshaya Shkola)

Nicolis G and Prigogine I 1977 *Self-Organisation in Non-Equilibrium Systems: From Dissipative Structures to Order through Fluctuations* (New York: Wiley)

Pokrovskii V N 2005 Extended thermodynamics in a discrete-system approach *Eur. J. Phys.* **26** 769–81

Prigogine I 1961 *Introduction to Thermodynamics of Irreversible Processes* 2nd edn (New York: Interscience)

Prigogine I and Defay R 1954 *Chemical thermodynamics* (London: Longman Green)

Shnol S 1997 *Geroi i zlodei rossiyskoy nauki [The Heroes and Villains of the Russian Science]* (Moscow: Kron)

Chapter 7

Dynamics of complex fluids

The aim of the chapter is to show how local equations of flow and deformation of material medium follow the principles of non-equilibrium thermodynamics. The first description was apparently laid out by Bernoulli and Euler in the eighteenth Century. Principles of thermodynamics for the continuous medium in the form of equations of balance were recorded in the beginning of the last century by Jaumann (1911) and Lohr (1917). The theory is formulated in the conventional form by Groot and Mazur (1962) and Landau and Lifshitz (1987). Further, we follow the article (Pokrovskii 1970), in which internal relaxation processes have been included in the general description of the problem. The principles of the theory of viscoelasticity are discussed, but no attempt is made to review available constitutive equations.

7.1 The equation of continuity and the law of conservation of momentum

A continuous medium is characterized by its mean density, a function of co-ordinates and time

$$\rho = \rho(\boldsymbol{x}, t).$$

The motion of a continuous medium is described by its velocity vector $\boldsymbol{v}$, which is a certain mean macroscopic velocity and has three components—functions of the co-ordinates and time—

$$v_i = v_i(\boldsymbol{x}, t), \quad i = 1, 2, 3.$$

The law of conservation of mass can be written in the form of the continuity equation

$$\frac{\partial \rho}{\partial t} + \operatorname{div} \rho \boldsymbol{v} = 0, \tag{7.1}$$

doi:10.1088/978-0-7503-3451-8ch7

where $\rho \boldsymbol{v}$ is the flux of mass density. Here and further on, the density of some quantity means the amount of this quantity in the volume unit of the medium.

The law of conservation of momentum can be written as

$$\frac{\partial(\rho v_i)}{\partial t} + \frac{\partial \Pi_{ik}}{\partial x_k} = \sigma_i$$

where $\Pi_{ik} = \rho v_i v_k - \sigma_{ik}$ is a tensor flux of the momentum density, which consists of the convective flux and the stress tensor; σ_i is density of the external forces that act on the fluid.

We can use the above relations to rewrite the law of conservation of momentum density in the form

$$\rho\left(\frac{\partial v_i}{\partial t} + v_j \frac{\partial v_i}{\partial x_j}\right) = \frac{\partial \sigma_{ik}}{\partial x_k} + \sigma_i. \tag{7.2}$$

For the observation times that is much greater than times of rotational relaxation of the structure elements of the liquid, internal and external rotation are balanced, and, thus, from equation (7.2) follows

$$\sigma_{ik} - \sigma_{ki} = -N_{ik}. \tag{7.3}$$

In this case, the stress tensor is non-symmetric, if there is an external force torque. At absence of external forces, the law of conservation of angular momentum follows from the law of conservation of momentum. We shall further assume, that the internal and external rotation are balanced and the stress tensor is symmetric.

7.2 The law of conservation of energy and balance of entropy

We assume that there are no internal sources of energy in the fluid, so that the change of the total energy density U is connected with fluxes through the surface of the volume. The law of the conservation of energy can be written in the form

$$\frac{\partial U}{\partial t} + \operatorname{div} \boldsymbol{q} = 0 \tag{7.4}$$

where $\boldsymbol{q}$ is the flux of energy density. The density of total energy U in equation (7.4) can be represented as a sum of the kinetic energy and the thermodynamic total energy of the resting volume

$$U = \frac{1}{2}\rho v^2 + U_0. \tag{7.5}$$

The law of the conservation of energy is also known as the first principle of thermodynamics.

To formulate the motion equation of a liquid, it is necessary also to use the second principle of thermodynamics (Lohr 1917), which can be written as the equation for the change of the entropy s for unit mass. The balance equation for entropy density has the form

$$\frac{\partial(\rho s)}{\partial t} + \operatorname{div}(\boldsymbol{v}\rho s + \boldsymbol{H}) = \Sigma$$

where $\boldsymbol{H}$ is the non-convective flux of entropy density, Σ is the non-negative quantity of emerging of entropy—entropy production. This equation can be rewritten in another form

$$\rho\left(\frac{\partial s}{\partial t} + v_i\frac{\partial s}{\partial x_i}\right) + \operatorname{div}\boldsymbol{H} = \Sigma. \tag{7.6}$$

Equation (7.6) is valid in the most general case[1].

So as there is a thermodynamic relation between entropy and internal energy, the unknown quantities $\boldsymbol{q}$, $\boldsymbol{H}$ in equations (7.4) and (7.6) can be connected with each other and one can also determine production of entropy Σ through other quantities.

Note that the volume, fixed in the deformable liquid, does not contain a certain amount of substance, and therefore it is convenient to consider the unit of mass of moving substance as a thermodynamic system to which one could apply the results of the second chapter. Referring to relation (2.17), we write down for the case in question the change in the energy of the unit mass of the liquid depending on the changes in the arguments

$$\mathrm{d}\varepsilon = T\,\mathrm{d}s - p\,\mathrm{d}v + \frac{1}{\rho}\,\Xi_\alpha\,\mathrm{d}\xi^\alpha, \tag{7.7}$$

where p is the pressure, v and s are the volume and entropy of the mass unit. As non-equilibrium characteristics, formula (7.7) contains internal variables ξ^α, where the index α is used to numerate the variable and its tensor indices, and the corresponding thermodynamic forces

$$\Xi_\alpha = \left(\frac{\partial U_0}{\partial \xi^\alpha}\right)_{s,\rho} = -T\left(\frac{\partial(\rho s)}{\partial \xi^\alpha}\right)_{T,\rho} > 0.$$

[1] For systems, which are locally in the state of equilibrium, there is only convective transfer of entropy. This is the case of the ideal fluid, for which

$$\frac{\partial s}{\partial t} + v_i\frac{\partial s}{\partial x_i} = 0.$$

The entropy arises in systems, which can be considered as systems that are locally in equilibrium. The increase of entropy can be connected with heat production in units of volume of fluid or, in other words, with the dissipation of energy Φ.

$$\rho\left(\frac{\partial s}{\partial t} + v_i\frac{\partial s}{\partial x_i}\right) = \frac{\Phi}{T}.$$

Non-equality can be written for the case when we cannot consider the system to be locally in equilibrium.

$$\rho\left(\frac{\partial s}{\partial t} + v_i\frac{\partial s}{\partial x_i}\right) > \frac{\Phi}{T}.$$

In this general case, equation (7.6) is applied.

Using the definition of the energy unit of volume as $U_0 = \rho\varepsilon$, we write the differential of this value based on the ratio (7.7)

$$\mathrm{d}U_0 = \rho T\ \mathrm{d}s + w\ \mathrm{d}\rho + \Xi_\alpha\ \mathrm{d}\xi^\alpha, \tag{7.8}$$

where the notation for the heat content of the mass unit is introduced $w = \varepsilon + p/\rho$. The quantities T, w and Ξ_α are functions of the variables s, ρ, ξ^α. In equilibrium, when there are no external influences, all internal variables $\xi^\alpha = 0$, while the values T and w take their equilibrium values. The external field affects the internal variables, which determine the state of the system[2].

Now, taking relations (7.5) and (7.8) into account, we are ready to write down the rate of change of the density of the total energy of the moving fluid

$$\frac{\partial U}{\partial t} = \rho v_i \frac{\partial v_i}{\partial t} + \rho T \frac{\partial s}{\partial t} + \left(w + \frac{v^2}{2}\right)\frac{\partial \rho}{\partial t} + \Xi_\alpha \frac{\partial \xi^\alpha}{\partial t}.$$

We can use equations (7.1), (7.2) and (7.6) to transform the above expression to the equation which has the form of the law of the conservation of energy

$$\begin{aligned} &\frac{\partial U}{\partial t} + \frac{\partial}{\partial x_k}\left[\rho v_k\left(w + \frac{v^2}{2}\right) - v_i(\sigma_{ik} + p\delta_{ik}) + TH_k\right] \\ &= T\Sigma - (\sigma_{ik} + p\delta_{ik})\nu_{ik} + H_i \nabla_i T + \frac{\mathrm{d}\xi^\alpha}{\mathrm{d}t}\Xi_\alpha \end{aligned} \tag{7.9}$$

where $\nu_{ik} = \frac{\partial v_i}{\partial x_k}$ is the tensor of velocity gradient.

Comparison of equations (7.4) and (7.9) determines the flux of energy and dissipation of energy, correspondingly,

$$q_k = \rho v_k\left(w + \frac{v^2}{2}\right) - v_i(\sigma_{ik} + p\delta_{ik}) + TH_k, \tag{7.10}$$

$$\Sigma = \frac{1}{T}\left((\sigma_{ik} + p\delta_{ik})\nu_{ik} - H_i \nabla_i T - \frac{\mathrm{d}\xi^\alpha}{\mathrm{d}t}\Xi_\alpha\right). \tag{7.11}$$

Expression (7.10) defines the thermodynamic fluxes $\sigma_{ik} + p\delta_{ik}$ and H_k. The first two terms on the right side of the expression (7.11) define entropy production as the product (possibly with some multipliers) of the thermodynamic forces ν_{ik} and $\nabla_i T$ and specified thermodynamic fluxes, which corresponds to the general

[2] In non-equilibrium situations, local states of the deformed system are described by some internal thermodynamic variables ξ^α, where the label α is used for the number of a variable and its tensor indices. All the equilibrium values of the internal variables are functions of two thermodynamic variables: for example, density and entropy

$$\xi_e^\alpha = \xi_e^\alpha(s, \rho).$$

The deviation of the thermodynamic system from the equilibrium state is described by the differences $\xi^\alpha - \xi_e^\alpha$ which are noted as ξ^α henceforth.

expression (4.10). This means that processes with velocity and temperature gradients, as internal variables, are considered here in the conditions, when the relaxation times of internal variables are negligible compared to the characteristic times of fluid movement. The last terms in the right-hand side of the relation (7.11) determine the direct contribution to entropy production by relaxation of internal variables, as represented by the general expression (2.16); the relaxation times are not assumed to be small in this case. The equation of relaxation of internal variables should be included directly in the scheme of consideration.

Note that the internal variables ξ^α are introduced formally in relation (7.7). However, the success of specific applications depends on the proper selection of internal variables for the case under study. Consideration of heuristic models usually helps to recognize which quantities describe the deviation of the system from the equilibrium state and can be used as internal variables.

7.3 Thermodynamic fluxes and relaxation processes

To move further, we have to remember that, according to the second equation from (4.3), the thermodynamic fluxes in a stationary state can be considered as linear functions of thermodynamic forces (to say nothing of the other thermodynamic variables), and evolution of internal variables with noticeable relaxation times can be determined by equation (3.4), so that, in the general case, we should consider the dependences

$$\begin{aligned} \sigma_{ij} + p\delta_{ij} &= f_{ij}(\nu_{js}, \nabla_l T, \xi^\gamma) \\ -H_i &= H_i(\nu_{js}, \nabla_l T, \xi^\gamma), \\ -\frac{d\xi^\alpha}{dt} &= g^\alpha(\nu_{js}, \nabla_l T, \xi^\gamma). \end{aligned} \tag{7.12}$$

We must pay special attention to the last relation in equation (7.12), which is the relaxation equation for the variable ξ^α. Note that the diffusion of internal variables, that is, the diffusion of structural elements in the inhomogeneous distribution of values of internal variables, is ignored here[3].

Dependences (7.12) can be considered in a quasi-linear form, which does not limit the generality of consideration. It is convenient to use notations for symmetric and antisymmetric tensors of the velocity gradients

$$\gamma_{ij} = \frac{1}{2}(\nu_{ij} + \nu_{ji}), \qquad \omega_{ij} = \frac{1}{2}(\nu_{ij} + \nu_{ji})$$

and divide the stress tensor in symmetric and antisymmetric parts, to write the fluxes as a quasi-linear function of the forces

[3] Otherwise, the terms of the form $\frac{\partial^2 \xi^\alpha}{\partial x_j \partial x_l}$ would have to be added to the argument set on the right side of the functions in equation (7.12). We will not discuss this situation further.

$$
\begin{aligned}
\frac{1}{2}(\sigma_{ik} + \sigma_{ki} + 2p\delta_{ik}) &= \eta_{ikjs}\gamma_{js} + K_{ikjs}\omega_{js} + L_{ikj}\nabla_j T + M_{ik\alpha}\xi_\alpha,\\
\frac{1}{2}(\sigma_{ik} - \sigma_{ki}) &= B_{ikjs}\gamma_{js} + N_{ikjs}\omega_{js} + R_{ikj}\nabla_j T + S_{ik\alpha}\xi_\alpha,\\
-H_i &= U_{ijs}\gamma_{js} + C_{ijs}\omega_{js} + A_{ij}\nabla_j T + G_{i\alpha}\xi_\alpha,\\
-\frac{d\xi^\alpha}{dt} &= V_{\alpha js}\gamma_{js} + D_{\alpha js}\omega_{js} + WG_{\alpha i}\nabla_i T + P_{\alpha\gamma}\xi_\gamma.
\end{aligned}
\tag{7.13}
$$

The matrix coefficients in equation (7.13) depend on the thermodynamic variables, which, in the case under discussion, are pressure p or density ρ (we can chose any of them, so as there exists an equation of state, connecting these variables), temperature T and internal variables ξ^α. The coefficients in equation (7.13) can be expanded into series near equilibrium values of internal variables. All matrices in relations (7.13) are assumed to depend on the internal variables, though, actually, for most applications, it is sufficient to consider the linear approximation of the theory.

The zero-order terms of expansions of the components of the matrices in a series of powers of the internal variables (the stationery processes considered) are connected due to the Onsager principle (see section 4.5) by some relations

$$
B^\circ{}_{jsik} = K^\circ{}_{ikjs}, \quad R^\circ{}_{ikj} = -C^\circ{}_{jik}, \quad U^\circ{}_{jik} = -L^\circ{}_{ikj}. \tag{7.14}
$$

The superscript $^\circ$ denotes that all arguments of the functions have equilibrium values.

In the simple case when all the internal variables are scalar quantities, the state of the system is isotropic, all the matrix coefficients in equation (7.13) are expressed in unit matrices, and the relations (7.13) take the simpler form, which can be easily written for every given set of internal variables.

In rheological terms, equations (7.12) and (7.13) make up a set of constitutive relations of the system. Together with equations (7.1), (7.2) and (7.10), they determine the equations of motion of the system. The presence of internal variables and equations to change of them are specific features of liquids, which we call complex liquids, discussed in this chapter. The actual properties of the liquid are determined by selecting the internal variables ξ^α. Relations (7.13) can be easily written for each given set of internal variables. The actual properties of fluid are determined by choice of the internal variables ξ^α. A set of internal variables were identified, for example, for dilute polymer solutions and for polymer melts, which is described in the monograph (Pokrovskii 2010).

7.4 The principle of relativity for slow motions

The form of the above-written relations (7.13) can be specified further by applying some requirements, which follow from the assumption that the motion of structural elements of medium does not change very rapidly, so that the following relation is valid

$$\frac{u\rho a}{\eta} \ll 1. \tag{7.15}$$

Here, a is a characteristic size of the structural element, ρ is density, which is approximately equal to 1 g cm^{-3}, η is an effective viscosity coefficient of the medium, which is 10^{-2}–10 P s, and u is a characteristic velocity of motion of the particle, which is not more than the mean thermal velocity $(T/m)^{1/2}$. It is easy to see that, at room temperature and with the above values of the parameters, condition (7.15) is valid, if $a \gg 10^{-7} - 10^{-6}$ cm.

As is well known, the equations of mechanics are covariant with Galileo transform. This can be also said about relations (7.12) and (7.13). In the case when the motions of the internal particles are slow (in the sense discussed above), one can state that a stronger principle is valid (Jaumann 1911). It says that all the processes run in the same way and, consequently, should be described by similar equations in all co-ordinate frames that are connected to each other by the transform

$$x_i = a_{ik}x'_k + c_i \tag{7.16}$$

where an orthogonal tensor a_{ik} and a vector c_i are arbitrary functions of time. In contrast to the Galileo principle, the above principle, which is also called the principle of material objectivity, is valid for the cases when the forces of inertia can be neglected.

Let us consider the restriction imposed on the form of the transfer equations by the discussed principle. It is easy to see that, when transformation (7.16) is applied to co-ordinates, the tensor of velocity gradients transforms as

$$\nu_{ik} = a_{il}a_{kj}\nu_{lj}' + \dot{a}_{il}a_{kl}.$$

The superscript point denotes differentiation with respect to time.

The symmetrical tensor of velocity gradient transforms as a tensor, which does not depend on time

$$\gamma_{ik} = a_{il}a_{kj}\gamma'_{lj}. \tag{7.17}$$

The antisymmetric tensor transforms in the following way

$$\omega_{ik} = a_{il}a_{kj}\omega'_{lj} + \dot{a}_{il}a_{kl}. \tag{7.18}$$

Let us now turn to the internal variables. We can consider that one of the internal variables is a tensor of arbitrary rank and transforms as the co-ordinates do, that is, contravariantly

$$\xi_{ik\ldots l} = a_{ij}a_{ks}\ldots a_{ln}\xi'_{js\ldots n}.$$

Differentiating the tensor with respect to time, we find that

$$\frac{d\xi_{ik\ldots l}}{dt} = \dot{a}_{ij}a_{pj}\xi_{pk\ldots l} + \dot{a}_{kj}a_{pj}\xi_{ip\ldots l} + \cdots + \dot{a}_{lj}a_{pj}\xi_{ik\ldots p}$$
$$+ a_{ij}a_{ks}\ldots a_{ln}\frac{d\xi'_{js\ldots n}}{dt}.$$

We can define the expression $\dot{a}_{il}a_{kl}$ from (7.18) to rewrite the last expression in the form

$$\frac{\mathrm{d}\xi_{ik\ldots l}}{\mathrm{d}t} - \omega_{ip}\,\xi_{pk\ldots l} - \omega_{km}\,\xi_{im\ldots l} - \cdots - \omega_{ln}\,\xi_{ik\ldots n}$$
$$= a_{ij}a_{ks}\ldots a_{ln}\left(\frac{\mathrm{d}\xi'_{is\ldots n}}{\mathrm{d}t} - \omega'_{jq}\,\xi'_{qs\ldots n} - \omega'_{sq}\,\xi'_{jq\ldots n} - \cdots - \omega'_{nq}\,\xi'_{js\ldots q}\right).$$

We can see that the combination

$$\frac{\mathrm{D}\xi_{ik\ldots l}}{\mathrm{D}t} = \frac{\mathrm{d}\xi_{ik\ldots l}}{\mathrm{d}t} - \omega_{ip}\,\xi_{pk\ldots l} - \omega_{km}\,\xi_{im\ldots l} - \cdots - \omega_{ln}\,\xi_{ik\ldots n} \tag{7.19}$$

transforms as a tensor, which is independent of time. Expression (7.19) is called the Jaumann derivative of tensor $\xi_{ik\ldots l}$ with respect to time.

There are plenty of covariant derivatives of the tensor $\xi_{ik\ldots l}$ among which the Jaumann derivative has the simplest form. Indeed, expressions (7.18) and (7.19) are followed by the relation

$$\dot{a}_{il}a_{kl} = \omega_{ik} + \kappa\gamma_{ik} - a_{is}a_{kj}\left(\omega'_{sj} + \kappa\gamma'_{sj}\right)$$

where κ is the arbitrary constant. We can use this relation to introduce derivatives, which are generalisations of equation (7.19).

Covariant tensors can be considered in a similar way.

7.5 Viscoelastic fluids

One can now return to the set of equations (7.12) and (7.13) and apply to them the principle of covariance that were discussed in the previous section. Equations (7.13) under the requirement of being covariant at transformation (7.16) can be written in the simpler form

$$\begin{aligned}
\frac{1}{2}(\sigma_{ik} + \sigma_{ki} + 2p\delta_{ik}) &= \eta_{ikjs}\gamma_{js} + L_{ikj}\nabla_j T + M_{ik\alpha}\,\xi_\alpha,\\
\frac{1}{2}(\sigma_{ik} - \sigma_{ki}) &= S_{ik\alpha}\xi_\alpha,\\
-H_i &= U_{ijs}\gamma_{js} + A_{ij}\nabla_j T + G_{i\alpha}\,\xi_\alpha,\\
-\frac{\mathrm{D}\xi^\alpha}{\mathrm{D}t} &= V_{\alpha js}\gamma_{js} + W_{\alpha i}\nabla_i T + P_{\alpha\gamma}\,\xi_\gamma,
\end{aligned} \tag{7.20}$$

where the Jaumann derivative is defined as

$$\frac{\mathrm{D}\xi^\alpha}{\mathrm{D}t} = \frac{\mathrm{d}\xi^\alpha}{\mathrm{d}t} + D_{\alpha js}\,\omega_{js}. \tag{7.21}$$

All matrices in relations (7.20) are assumed to depend on the internal variables, the matrix $D_{\alpha js}$ can be determined for every given tensor ξ^α, by comparing equation (7.21) with definition (7.19).

The fourth line in the set (7.20) determines a set of relaxation equations for variables ξ^α that are responsible for viscoelastic behaviour of the system. If the relaxation processes are absent (all $\xi^\alpha = 0$), equations (7.20) turn into fundamental equations for the viscous fluid. One can see that the equations of motion for a viscoelastic fluid can always be written, when a set of internal relaxation variables is given, however, a set of internal variables cannot be determined in the frame of phenomenological theory and equations (7.20) cannot be specified any more without extra assumptions. The most known example of viscoelastic fluids are polymers and their solutions. It was discovered (Pokrovskii 2010) that one needs two sets of internal variables to formulate fundamental equations for this case correctly. To describe viscoelasticity of weakly entangled polymer systems, one needs a set of conformational variables x_{ik}^α that characterize the deviations of the form and size of macromolecular coils from the equilibrium values. To describe behaviour of strongly entangled systems, one needs, apart from the set of conformational variables, the other set of orientational variables u_{ik}^α which are connected with the mean orientation of the segments of the macromolecules.

As an example, we shall consider a simpler case of the isothermal motion of a liquid without external volume forces and without external volume force torque. To consider this case in more detail, we specify equations (7.20) and (7.21) for a single internal variable—the symmetric tensor of the second rank ξ_{js}. We do not know the dependences of the matrices in equations (7.20) on the internal variable, so that it is convenient to use the notions

$$\bar{\sigma}_{ik}(\boldsymbol{\xi}) = M_{ikjs}(\boldsymbol{\xi})\,\xi_{js}, \qquad \phi_{ij}(\boldsymbol{\xi}) = P_{ijls}(\boldsymbol{\xi})\,\xi_{ls}$$

and write the constitutive relations as follows

$$\begin{aligned} \sigma_{ik} + p\delta_{ik} &= \eta_{ikjs}(\boldsymbol{\xi})\,\gamma_{js} + \bar{\sigma}_{ik}(\boldsymbol{\xi}), \\ -\frac{\mathrm{D}\xi_{ij}}{\mathrm{D}t} &= V_{ijls}(\boldsymbol{\xi})\,\gamma_{ls} + \phi_{ij}(\boldsymbol{\xi}). \end{aligned} \tag{7.22}$$

On this step, one has to attract empirical evidence and to apply some simplifications. For the polymer fluid as an example, the first term on the right-hand side of the first equation from set (7.22) can be neglected, and the matrix in the first term on the right-hand side of the second equation can be presented in the linear form

$$V_{ijls} = -\frac{1}{2}\kappa(\xi_{il}\delta_{js} + \xi_{js}\delta_{il} + \xi_{is}\delta_{jl} + \xi_{jl}\delta_{is}) \tag{7.23}$$

where κ is a numerical parameter, usually taken as ± 1 or 0.

The second terms on the right-hand side of equations (7.22) can be presented as a general function of the tensor internal variable ξ_{il}

$$\begin{aligned} \bar{\sigma}_{ik} &= \sigma_0\delta_{ik} + \sigma_1\xi_{ik} + \sigma_2\xi_{il}\xi_{lj}, \\ \phi_{ik} &= \phi_0\delta_{ik} + \phi_1\xi_{ik} + \phi_2\xi_{il}\xi_{lk}, \end{aligned} \tag{7.24}$$

where the coefficients σ_i and ϕ_i $(i = 0, 1, 2)$ are functions of the three invariants of the tensor ξ_{il}

$$I_1 = \sum_{i=1}^{3} \xi_{ii}, \quad I_2 = \frac{1}{2}\sum_{i,j} (\xi_{ij}\xi_{ji} - \xi_{ii}\xi_{jj}), \quad I_3 = |\xi_{ij}|. \tag{7.25}$$

The relations (7.22)–(7.25) make up a general form for a non-linear single-mode constitutive relation. This form embraces many empirical constitutive equations for viscoelastic polymer fluids. One can look, for an example, at equations (43) and (44) in the article (Pyshnograi *et al* 2009).

7.6 Different forms of constitutive relation

The constitutive relations for viscoelastic fluids that we have discussed in the previous sections contain some relaxation equations for the internal variables which ought to be considered as independent variables in the system of equations for the dynamics of liquids. However, in earlier times, beginning with the works by Lord Kelvin (Thomson 1865), Maxwell (1867) and Boltzmann (1874), the constitutive relations for a viscoelastic liquids were formulated as relations between the stress tensor and the kinetic characteristics of the deformed state of the medium without mentioning internal variables. In this section, we shall show that the constitutive relation with internal variables is followed by two types of constitutive relations which do not include internal variables.

For the sake of simplicity, we shall consider the simplest set of equations

$$\sigma_{ik} + p\delta_{ik} = 3\frac{\eta}{\tau}\left(\xi_{ik} - \frac{1}{3}\delta_{ik}\right), \tag{7.26}$$

$$\frac{\mathrm{d}\xi_{ik}}{\mathrm{d}t} - \nu_{ij}\xi_{jk} - \nu_{kj}\xi_{ji} = -\frac{1}{\tau}\left(\xi_{ik} - \frac{1}{3}\delta_{ik}\right) \tag{7.27}$$

where the coefficient of viscosity η and the time of relaxation τ are functions of the invariants (7.25) of the internal tensor variable ξ_{ik}. By excluding the tensor ξ_{ij} from the set of equations (7.26) and (7.27), we can obtain a relation between the stress tensor and the velocity gradient tensor. This can be done in two different ways.

Firstly, from equation (7.26), we can define the tensor ξ_{ij} which can be inserted into the second equation of (7.27). As a result, we obtain a differential equation for the extra stresses

$$\frac{\mathrm{d}\tau_{ik}}{\mathrm{d}t} - \nu_{ij}\tau_{jk} - \nu_{kj}\tau_{ji} = -\frac{1}{\tau}(\tau_{ik} - 2\eta\gamma_{ik}), \quad \tau_{ik} = \sigma_{ik} + p\delta_{ik}. \tag{7.28}$$

The quantities τ and η in equation (7.28) depend on the invariants of the tensor τ_{ik} in accordance with equation (7.26). We ought to note that the behaviour of a non-linear viscoelastic liquid in a non-steady state would be different, if a dependence of the material parameters τ and η on the tensor velocity gradients or on the stress tensor is assumed.

In the linear case, when the relaxation time τ and coefficient of shear viscosity η are constant, the relation (7.28) coincides with that derived by Maxwell (1867)

$$\frac{\mathrm{d}(\sigma_{ik} + p\delta_{ik})}{\mathrm{d}t} + \frac{1}{\tau}(\sigma_{ik} + p\delta_{ik}) = 2\frac{\eta}{\tau}\gamma_{ik}. \tag{7.29}$$

A refined exposition of formulation can be found in the book of Landau and Lifshitz (1988, p 36).

On the other hand, we can imagine that a solution of equation (7.27) can be found. Below, for the example, the solution is written for uniform flow with accuracy up to the second-order terms with respect to the velocity gradient

$$\begin{aligned}\xi_{ik} = \frac{1}{3}\delta_{ik} &+ \int_0^\infty \exp\left(-\frac{s}{\tau}\right)\gamma_{ij}(t-s)\mathrm{d}s \\ &+ \int_0^\infty \exp\left(-\frac{s}{\tau}\right)\int_0^\infty \exp\left(-\frac{u}{\tau}\right) \\ &\times\left[\nu_{ij}(t-s)\gamma_{jk}(t-s-u) - \nu_{kj}(t-s)\gamma_{ji}(t-s-u)\right]\mathrm{d}u\,\mathrm{d}s.\end{aligned}$$

Then, the solutions should be inserted into equation (7.26), which determines the stress tensor as a function of the tensor of the velocity gradient in the previous moments of time. The linear term has the form

$$\sigma_{ij} = -p\delta_{ij} + 2\frac{\eta}{\tau}\int_0^\infty \exp\left(-\frac{s}{\tau}\right)\gamma_{ik}(t-s)\mathrm{d}s. \tag{7.30}$$

A generalization of equation (7.30) for the case of many relaxation processes can easily be found. In the simplest case of uniform motion one has

$$\sigma_{ik} = -p\delta_{ik} + 2\int_0^\infty \eta(s)\gamma_{ik}(t-s)\mathrm{d}s. \tag{7.31}$$

The memory function $\eta(s)$ can be calculated if a set of internal variables is given. It should be noted that the first person apparently to use functional relations to describe the behavior of viscoelastic materials was Boltzmann (1874). In the general case, the stress tensor can be written as a functional of the history of velocity gradients or displacement gradients and laid in the basis of the phenomenological theory of so-called simple materials (Coleman and Noll 1961). To put the theory into practice, the function should be, for example, represented by an expansion into a series of repeated integrals, so that, in the simplest case, one has the first-order constitutive relation (7.31).

One can see that there are different forms for the representation of the constitutive relation for a viscoelastic liquid. Of course, we ought to say that all the types of constitutive relation that were discussed in this section are equivalent. We can use any of them to describe the flow of viscoelastic liquids. However, the description of the flow of the liquid in terms of the internal variables allows one to use additional information, if it is available, about microstructure of the material, and, in fact, appears to be the simplest one for derivation and calculation. We believe that the forms, which include the internal variables, reflect a deeper penetration into the mechanisms of the viscoelastic behaviour of materials. From this point of view, all the representations of deformed material can be unified and classified.

References

Boltzmann L 1874 Zur Theorie der Elastischen Nachwirkung *Sitzungsber. Akad. Wiss. Wien Math. Naturwiss.* **70** 275–300

Coleman B D and Noll W 1961 Foundations of Linear Viscoelasticity *Rev. Mod. Phys.* **33** 239–49

Groot de S R and Mazur P 1962 *Non-equilibrium Thermodynamics* (Amsterdam: North-Holland)

Jaumann G 1911 Geschlossenes System physicalischer und chemischer Differentialgesetze *Sitzungber. Akad. Wiss. Math.-Nat. Kl. Wien Abt. 2a* **120** 385–530

Landau L D and Lifshitz E M 1987 *Fluid Mechanics* 2nd edn (Oxford: Pergamon)

Landau L D and Lifshitz E M 1988 *Theory of Elasticity* 3rd edn (Oxford: Pergamon)

Lohr E 1917 Entropieprinzip und geschlossenes Gleichungsystem *Denkschr. Akad. Wiss. Math.-Naturwiss. Kl.* **93** 339–421

Maxwell J C 1867 On the dynamical theory of gases *Trans. R. Soc.* **157** 49–88

Pokrovskii V N 1970 Equations of motion of viscoelastic systems as derived from the conservation laws and the phenomenological theory of non-equilibrium processes *Polym. Mech.* **6** 693–702

Pokrovskii V N 2010 *The Mesoscopic Theory of Polymer Dynamics* (Dordrecht: Springer)

Pyshnograi G V, Gusev A S and Pokrovskii V N 2009 Constitutive equations for weakly entangled linear polymers *J. Non-Newtonian Fluid Mech.* **164** 17–28

Thomson W W 1865 On the elasticity and viscosity of metals *Proc. R. Soc.* **14** 289–97

IOP Publishing

Thermodynamics of Complex Systems
Principles and applications
Vladimir N Pokrovskii

Chapter 8

Creation of complexity

During transition of the open thermodynamic system from one state to another under abundance of available energy and matter, entropy of the system could decrease, which is connected with increasing of complexity. It appears to be possible due to a natural mechanism of delayed relaxation of internal variables. Apart from that, there is another mechanism of creation of complexity, described with the help of the notion of information, which, alongside with the concept of entropy, is used during the description of complex systems. The fundamental foundation of possibility of transition to the higher forms of organization is a many-particle composition of the thermodynamic system, which is discussed in the second section of the chapter. Eventually, we shall be able to describe the practical methods of the creation of artificial complexity.

8.1 Complexity and information

8.1.1 Index of complexity

Considering the thermodynamic system in non-equilibrium state T, $\boldsymbol{x}$, $\boldsymbol{\xi}$, one can notice that, in addition to the various thermodynamic characteristics of the system, the concept of *complexity*, which is intuitively associated with the set of internal variables $\boldsymbol{\xi}$, can be conveniently used. We believe that the values and mutual interactions of internal variables completely determine the structure of the system; it is not without reason to call the internal variables *variables of complexity* or *variables of order*. The notion of *complexity*, as a characteristic of the system, in line with a notion of entropy of the system S, ought to be introduced and can be defined as

$$C(T, \boldsymbol{x}, \boldsymbol{\xi}) = S_0(T, \boldsymbol{x}) - S(T, \boldsymbol{x}, \boldsymbol{\xi}). \tag{8.1}$$

The index of complexity can be considered as characteristic of the structure of the system; the quantity is connected with the change of entropy of the system: by virtue of equation (8.1), increase or decrease in number and values of internal variables

doi:10.1088/978-0-7503-3451-8ch8

provokes decrease or increase in entropy. In the equilibrium situation the index of complexity is equal to zero.

Let us note that the quantity (8.1) is usually meant, when one uses notions of *bound information* (Brillouin 1956), or *information content* (Trincher 1965), or just simply *information of the system*. Instead of these terms, it is more convenient to use the word *complexity* to emphasize the difference between a thermodynamic term and the concept of *information*, used in different meanings, in different situations, and in relation to different systems. There is a need to look at this concept more closely.

8.1.2 Concept of information

To recall the definition of the concept of *information*, applicable to physical problems, we shall begin with the notion of a measure of uncertainty. We shall imagine that the problem with some set of possible answers is considered, and there is no special knowledge of the right answer. For the characteristic of the situation *a measure of uncertainty* can be introduced (Shannon 1948). In the case when there are n possible various answers with probabilities p_i, the measure of uncertainty is recorded as

$$-\sum_{i}^{n} p_i \log p_i, \quad 0 \leqslant p_i \leqslant 1. \tag{8.2}$$

Every term of this sum and the entire sum are always positive due to admissible values of probabilities of events.

If it somehow becomes known that the number of possible answers in the considered problem can be reduced, or even that the unique possible answer can be known, we say that we have received some information on the situation. It is essential that, in the considered situation, value of probabilities of realization of events, which now become equal to p_i' is specified. Then the measure of uncertainty of a new state looks like

$$-\sum_{i}^{n} p_i' \log p_i', \quad 0 \leqslant p_i \leqslant p_i' \leqslant 1.$$

The measure of uncertainty decreases (more precisely, does not increase) at reception of the information. At authentic knowledge of an outcome, the measure of uncertainty of a new condition is equal to zero.

To characterize transition from the uncertain state to a more certain one, we introduce the quantitative measure of the information

$$I = -\sum_{i}^{n} p_i \log p_i + \sum_{i}^{n} p_i' \log p_i'. \tag{8.3}$$

Thus, at definition of concept of *information* two states of the same situations with various degrees of uncertainty are compared. The measure of information of the message on the certain outcome is equal to the measure of uncertainty of the initial situation, so in this case

$$I = -\sum_{i}^{n} p_i \log p_i. \qquad (8.4)$$

Transition from one state of awareness to another occurs at reception of a message, which carries the information, thus it is supposed, that there is a certain sender and, at least one addressee of the message. Simply speaking, the information is a content of the message that is received and realized. But, in general, the information is the result of processing, management and structuring of the data in such a way that adds something to the knowledge of a person, who is obtaining information. The information always is a message about something (size of parameter, occurrence of a case $\cdots$) and a characteristic of process.

8.2 Statistical interpretation of ideal gas

The microscopic, statistical description of thermodynamic systems goes back to works of Maxwell (1871) and Boltzmann (1877), who considered ideal gas, as an assembly of N particles, each of which obeys the laws of classical mechanics, according to which state of any particle can be fixed by values of coordinates and momentum r_1, r_2, r_3, p_1, p_2, p_3. Each particle can be ascribed a scalar characteristic—energy, which, generally speaking, is a function of coordinates and momenta $\varepsilon = \varepsilon(\boldsymbol{r}, \boldsymbol{p})$. The energy of the particle changes due to the momentary collisions of the particles, but, nevertheless, in every moment of time the particles of the ensemble can be considered as non-interacting.

On the example of an ideal gas, it was shown that the properties of the system are not reduced to the sum of the characteristics of individual components, but new characteristics of the system as a whole appear. For thermodynamic systems, it is the entropy that controls the macroscopic changes in the system. It was a revolution in describing the properties of thermodynamic systems, to establish the molecular interpretation of entropy (Sommerfeld 1952).

8.2.1 The phase space

The set of all admissible values of coordinates and impulses r_1, r_2, r_3, p_1, p_2, p_3 forms *the phase space of a particle*, which can be unlimited, if coordinates or/and momenta of the particles accept unlimited values. It is possible to assume, that some restrictions are imposed on the system. For example, the system could be located in a certain volume or energy of the system should remain a constant. In this case the phase space appears limited.

The phase space of a particle can be divided into cells. Each cell can be attributed with the index $\boldsymbol{r}, \boldsymbol{p}$ that specifies the cell, whose volume in multivariate space is located between values of variables r_1, r_2, r_3, p_1, p_2, p_3 and $r_1 + \Delta r_1$, $r_2 + \Delta r_2$, $r_3 + \Delta r_3$, $p_1 + \Delta p_1$, $p_2 + \Delta p_2$, $p_3 + \Delta p_3$. We assume, that each cell has volume $\Delta \boldsymbol{r} \Delta \boldsymbol{p}$.

In every moment of time every particle has definite values of coordinates and momenta, so that the numbers of particles in every cell are given; one can label $\Delta_{\boldsymbol{r},\boldsymbol{p}} N$ as number of particles in a cell of the standard size with coordinates $\boldsymbol{r}$, $\boldsymbol{p}$, for example.

In a similar way, one can consider $\Delta_r N$, the number of particles with either values of momenta, but with coordinates between r_1, r_2, r_3 and $r_1 + \Delta r_1, r_2 + \Delta r_2, r_3 + \Delta r_3$, or $\Delta_p N$, a number of particles with either values of coordinates, but with momenta between p_1, p_2, p_3 and $p_1 + \Delta p_1, p_2 + \Delta p_2, p_3 + \Delta p_3$. To determine these quantities, one has apparently to sum all numbers at given values of co-ordinate or momentum

$$\Delta_r N = \sum_p \Delta_{r,p} N, \quad \Delta_p N = \sum_r \Delta_{r,p} N. \tag{8.5}$$

If numbers of particles in each cell are set, apparently, the total number of particles and total energy of the system can be recorded as

$$N = \sum_{r,p} \Delta_{r,p} N, \quad U = \sum_{r,p} \varepsilon(\boldsymbol{r}, \boldsymbol{p}) \Delta_{r,p} N, \tag{8.6}$$

where energy of a separate particle $\varepsilon = \varepsilon(\boldsymbol{r}, \boldsymbol{p})$, generally speaking, is a function of coordinates and momentum. The quantities (8.6) are constant for an isolated system. For a closed system, energy is not a constant; for an open system, energy and number of particles can change.

Considering the isolated system, we shall notice, that, at the given total number of particles N in the system, distribution of particles over cells can be various. In each given point in time the system has the certain distribution of particles over cells, about which one speaks of *a microstate* of the system, corresponding to *macrostate* with preset values N and U. However, as we already know, apart from these quantities, one needs entropy for the total macroscopic description of a thermodynamic system. There appears a question of microscopic interpretation of entropy.

8.2.2 The number of microstates of the system

Let us consider the closed system with the total number of particles N, occupying the given volume V. To count the number of microstate in phase state, it is convenient to enumerate all cells and consider distribution of total number of N particles over the cells. One assumes that a certain number of particles occupy each cell, so that one has a set of *numbers of filling*

$$N_1, N_2, N_3, \dots, N_\alpha, \dots.$$

The total number of particles of the system is set

$$\sum_\alpha N_\alpha = N. \tag{8.7}$$

For visualization, it is possible to represent the distribution of particles over the spatial volume broken into the cells.

The number of possible distributions of particles over the cells (number of microstates) for the given number of particles N can be estimated as

$$\Omega(N_1, N_2, N_3, \dots, N_\alpha, \dots) = \frac{N!}{N_1!N_2!N_3!\dots, N_\alpha!\dots}. \tag{8.8}$$

To find out the character of dependence of the number of microstates on the distribution of particles over cells, we shall consider variation of one of the numbers, for example, N_α. As the total number of particles, on assumption, is invariable, the change of one of the numbers of filling is connected with a change of the other, suppose, with the number N_γ. At variation of distribution of particles over cells, the energy of the system and total number of particles do not change, so that the same macrostate with preset values N and E corresponds to the microstate with new numbers of filling, and the number of microstates can be defined as

$$\Omega(N_1, N_2, \dots N_\alpha + \Delta N, \dots N_\gamma - \Delta N, \dots) = \frac{N!}{N_1!\dots(N_\alpha + \Delta N)!\dots(N_\gamma - \Delta N)!\dots}.$$

When the moving of an arbitrary number ΔN of particles from one cell to another is occurring, the apparent relations between the factorials of the numbers of particles in the two situations can be recorded as

$$(N_\alpha + \Delta N)! = (N_\alpha + \Delta N)(N_\alpha + \Delta N - 1)\dots(N_\alpha + 1)\, N_\alpha!$$

$$(N_\gamma - \Delta N)! = \frac{N_\gamma!}{N_\gamma(N_\gamma - 1)\dots(N_\gamma - \Delta N + 2)(N_\gamma - \Delta N + 1)}.$$

The above-written formulas allow one to establish a connection between numbers of microstates when moving of several particles from one cell into another

$$\Omega(N_1, N_2, \dots N_\alpha + \Delta N, \dots N_\gamma - \Delta N, \dots) = \frac{N_\gamma(N_\gamma - 1)\dots(N_\gamma - \Delta N + 1)}{(N_\alpha + \Delta N)(N_\alpha + \Delta N - 1)\dots(N_\alpha + 1)}\,\Omega(N_1, N_2, N_3, \dots, N_\alpha, \dots). \tag{8.9}$$

At arbitrary valucs of numbers of filling N_α and N_γ, the multiplier in the last equation could be both greater or less than unit, however one could notice, that at $N_\alpha = N_\gamma$ the multiplier accepts values of less than unit at any ΔN. It means, that considering the state, in which all cells contain the identical number of particles, the moving of the particles from one cell into another, leads to a decrease in the number of microstates. Simple reasoning convinces us that the distribution with identical numbers of particles in the calls, is a unique state with the greatest value of number of microstates. Any moving (not trivial) of the particles between cells leads to reduction of the number of microstates.

Thus, among a huge number Ω of microstates or microrealisation of the macrostate, there is the only state with equal numbers, but, due to the huge number of particles, there are a huge number of different states.

8.2.3 Thermodynamic probability

The number of microstates Ω has appeared to be an important characteristic of the system. Though in every given moment of time the system is in the definite microstate, one knows that there are Ω microstates corresponding to the same macrostate. There is a huge number of microrealisations of the only macrostate. The system has a definite number of the total amount of particles, N, but the distribution of the particles over the cells can be different.

Continuing consideration of the system with a certain number of particles N within a volume V, it is possible to assume, that each microstate of the system, can be realized without any preference, so that the probability of the system to be in a state with the fixed numbers of filling of each cell can be recorded as

$$W(N_1, N_2, N_3, \ldots, N_\alpha, \ldots) = \frac{\Omega(N_1, N_2, N_3, \ldots, N_\alpha, \ldots)}{\Omega(N, V)}, \tag{8.10}$$

where $\Omega(N, V)$ is the total of microstates of the system, which, naturally, is calculated as the sum of the numbers of all possible microstates

$$\Omega(N, V) = \sum_{N_1, N_2, \ldots} \Omega(N_1, N_2, N_3, \ldots, N_\alpha, \ldots). \tag{8.11}$$

The total number of microstates is a characteristic of the system.

Function (8.10) allows calculation of the mean number of particles in a cell

$$\bar{N}_\gamma(N, V) = \sum_{N_1, N_2, \ldots} N_\gamma W(N_1, N_2, N_3, \ldots, N_\alpha, \ldots) \tag{8.12}$$

and the mean value of energy of the system

$$\begin{aligned} U(N, V) &= \sum_{N_1, N_2, \ldots} (\varepsilon_1 N_1 + \varepsilon_2 N_2 + \cdots + \varepsilon_\alpha N_\alpha) \\ &\quad \times W(N_1, N_2, N_3, \ldots, N_\alpha, \ldots) \\ &= \sum_{\alpha=1,2,\ldots} \varepsilon_\alpha \bar{N}_\alpha. \end{aligned} \tag{8.13}$$

For the isolated system, energy does not depend on time. For the closed system, energy is not a constant; for the open system, energy and number of a particle are not constant.

8.2.4 Entropy and temperature

One of the main characteristics of the thermodynamic system is entropy S, about which it is known, that it has the greatest value in the equilibrium state. Apparently, it was Boltzmann who first had the idea to compare entropy with the number of microstates of the system considered the dependence

$$S = S(\Omega).$$

The number of microstates was assumed to correspond to entropy, while the state with identical number of particles in cells (equally-distributed) corresponds to the equilibrium thermodynamic state.

Entropy of the system is an additive characteristic, so that, at association of two systems, the total number of microstates of the compound system is the product of numbers of microstates of the united systems, Boltzmann guessed that entropy of the system is a logarithmic function of number of microstates, and, at the given distribution of particles over cells, can be recorded as

$$S(N_1, N_2, N_3, \dots, N_\alpha, \dots) = \ln \Omega(N_1, N_2, N_3, \dots, N_\alpha, \dots). \tag{8.14}$$

Entropy can be considered as a non-dimensional quantity, and a possible factor of proportionality is omitted, as insignificant for the definition of entropy (Landau and Lifshitz 1986).

The mean value of entropy for the system with fixed macrostate is determined with the help of function of probability (8.10)

$$\begin{aligned} S(N, V) = \sum_{N_1, N_2, \dots} & \ln \Omega(N_1, N_2, \dots, N_\alpha, \dots) \\ & \times W(N_1, N_2, \dots, N_\alpha, \dots). \end{aligned} \tag{8.15}$$

This expression can be presented, by virtue of relations (8.10) and (8.11), as

$$\begin{aligned} S(N, V) = & \ln \Omega(N, V) \\ & + \sum_{N_1, N_2, \dots} \ln W(N_1, N_2, \dots, N_\alpha, \dots) \\ & \times W(N_1, N_2, \dots, N_\alpha, \dots). \end{aligned} \tag{8.16}$$

The expressions for entropy (8.15) and (8.16) are valid for both equilibrium, and non-equilibrium situations. For the equilibrium state with exact number of filling, the distribution function turns in product of delta-functions, and it gives

$$S_0(N, V) = \ln \Omega(N, V). \tag{8.17}$$

The numbers of microstates in a non-equilibrium and the equilibrium states are connected by the formula of a type (8.9), so that it is possible to present the formula for entropy (8.15) in the form

$$\begin{aligned} S(N, V, \boldsymbol{\xi}) = & S_0(N, V) \\ & + \sum_{N_1, N_2, \dots} \ln f(\Delta N_1, \Delta N_2, \dots \Delta N_\alpha, \dots) \\ & \times W(N_1, N_2, N_3, \dots, N_\alpha, \dots), \end{aligned} \tag{8.18}$$

where ΔN_α represent a deviation of precise numbers of filling from their equilibrium values. The internal variables $\boldsymbol{\xi}$ can be interpreted also as the deviations of the mean values of numbers of filling from their equilibrium values

$$\xi_\gamma = \bar{N}_\gamma - \bar{N}_\gamma^0.$$

Apparently, at the preset numbers of particles, the deviations of the numbers from equilibrium values are not independent, and the number of internal variables does not exceed the number of cells, and, as a rule, is much less than this number.

Thus, the expression for entropy of the system can be presented as a function of number of particles, volume and a set of internal variables. Now, according to equation (2.29), we can define absolute or reference temperature that is measured in power units,

$$T = \frac{\partial U}{\partial S}. \tag{8.19}$$

This expression is valid for both the equilibrium and non-equilibrium situation. Now the other thermodynamic functions can be calculated.

8.3 Does a thermodynamic system contain information?

The molecular composition of the thermodynamic system provides a multitude of various forms of the higher organization of matter, the simplest characteristic of which appears to be the index of complexity (8.1) that, due to equations (8.16) and (8.17), can be represented as

$$\begin{aligned} C(T, V, \boldsymbol{\xi}) &= S_0(T, \boldsymbol{x}) - S(T, \boldsymbol{x}, \boldsymbol{\xi}) \\ &= - \sum_{N_1, N_2, \dots} \ln W(N_1, N_2, \dots, N_\alpha, \dots) \\ &\quad \times W(N_1, N_2, \dots, N_\alpha, \dots), \end{aligned} \tag{8.20}$$

where $W(N_1, N_2, \dots, N_\alpha, \dots)$ represents the probability that the system is in a microstate with fixed filling numbers.

Note that the expression (8.20) formally coincides with the expression (8.4) for the measure of information, and we can, following this correspondence, interpret the complexity index (8.20) as the information that the researcher must get in order to go from describing the situation of the ignorance of microstate of the system to a full description when the numbers of filling of all cells of the phase space are known. As can be seen from (8.20), for the system in equilibrium, this information is zero in accordance with our idea of the equidistribution of particles across cells. In a non-equilibrium situation with the set values of internal variables $\boldsymbol{\xi}$, the information required to identify the microstate is greater when the deviation from equilibrium is greater. Note that we use the concept of information to think about our knowledge; we can compare two of our representations of a thermodynamic system and judge what information we need to get in order to move from one representation to the other.

Formal identity of the expression (8.20) as a characteristic of the thermodynamic system and equation (8.4) for the measure of information for the transition from a situation of uncertainty to a totally defined situation provoked[1] the identification of the concepts of information and entropy, which can be justified. The complexity index represents a characteristic of a thermodynamic system, and therefore we could conclude that the information now called *bound information* (Brillouin 1956) is also objective; each thermodynamic system can be assigned information that is identical to the complexity index. In the absence of the sender and recipient of the message, difficulties arise in any interpretation of the bound information, however, we can now say that external influences have an informational content: when the transition from the equilibrium state to a non-equilibrium state occurs and some internal variables ξ_1, ξ_1, ... appear, the system receives information. We conclude that the *information received by the system* is equal to a decrease in the entropy of the system (Brillouin 1956, chapter 12), that is, in other words, equal to a change in the bound information or complexity. Unfortunately, the reasoning in this paragraph does not advance us further in understanding the situation, but it cannot be ignored, since the question of how much information is contained in a thermodynamic system is often raised and discussed, but in fact it is a question of evaluating the entropy of the system.

The question of how much information in a system, generally speaking, is meaningless, and it is very unfortunate to have such a confusion of concepts of entropy and information, which we owe, in many respects, to the French physicist Brillouin (1956), who included the concept of information in the discipline of thermodynamics. To prevent possible misunderstandings, we will use the term *complexity* when discussing the features of physical objects, and leaving the meaning, which is explained in section 8.1.2, to the term *information.* To describe the situation correctly, we need to know how to estimate the amount of information in the impacts that are received from the closest agents. In the next section, we consider an example that allows us to evaluate directly the change in entropy and the amount of information received by the system during the transition from the equilibrium to a non-equilibrium state, which allows us to discuss the relationship between the concepts of entropy and information when creating a non-equilibrium state.

8.4 Production of complexity

Humans live in an artificial environment, which, in the same way as the natural environment, appears now to be an object of investigation by science (Simon 1996). Being interested in processes of production of the artificial things around us, we

[1] It was told (Tribus and McIrvine 1971), that Shannon did not know how to dub the measure introduced by him $-\sum p_i \log p_i$, and he asked for advice from von Neumann, who said: 'You should call it 'entropy' for two reasons: first, the function is already in use in thermodynamics under that name; second, and more importantly, most people don't know what entropy really is, and if you use that word entropy you will win every time'. Probably, von Neumann was only joking, anyway, his words are similar to a joke, and we can believe that von Neumann knew what entropy was, but Shannon accepted the advice sincerely and named the measure of information as entropy, having created the basis for terminological confusion.

discover everywhere thermodynamic cycles of production, which are developed to transform 'wild' forms of substances into 'useful' forms (houses, food, clothing, cars and so on).

8.4.1 Elementary production cycle

The performance of thermic machines, which are designed to transform heat into mechanical work, is described by thermodynamics by the means of thermodynamic cycles. Thermodynamics itself has emerged from studying this kind of process (Kondepudi and Prigogine 1999). But humans deal also with cycles of other kind, the aim of which is to create a non-equilibrium situation. The problem arises to analyze the process of production of artificial things and answer the question (that was interesting for Maxwell still), how a new non-equilibrium situation can be created. The problem was vividly discussed over decades. Further on, a simple example to demonstrate what one needs in order to create a non-equilibrium state of matter artificially will be considered.

Let us consider a system of $2N$ particles (ideal gas) in a container consisting of two compartments of volume V each, as shown in figure 8.1. There are some devices, which allow the compartments to be connected or isolated (let us call this operation A) and the volume of the second department to be diminished or restored to the previous volume (operation B).

Let us assume that in an initial state each compartment has volume V and the compartments are connected with each other, while gas is in an equilibrium state, so that each compartment has on average N particles. We consider isothermal processes consisting of several elementary operations, while every operation is fulfilled in a reversible manner. One can imagine that a deliberate sequence of operations can be applied to the system. After one has performed the sequence: B—decreasing of volume 2 in ΔV, A—isolating of the compartments, B—increasing of volume 2 in ΔV, and A—connecting the compartments, the number of particles in each compartment can be found to be

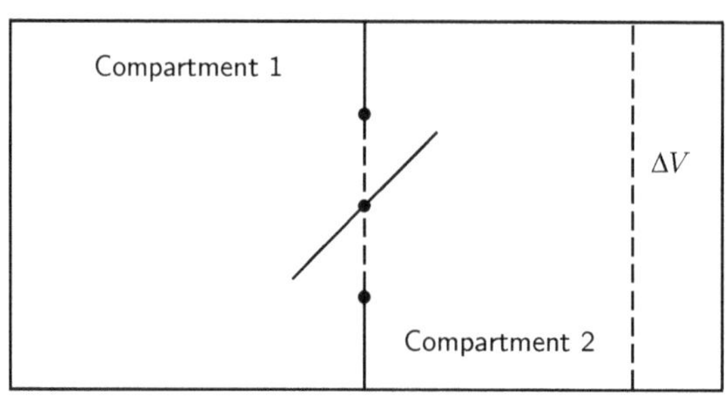

Figure 8.1. The scheme of the production container. The production devices can perform two operations: operation A allows the compartments to be connected or isolated and operation B allows the volume of the second department to be diminished or restored. Starting from the state of connected compartments, the sequence of operations BABA leads to creation of a non-equilibrium state of the working body.

$$N_1 = N(1 + \xi), \quad N_2 = N(1 - \xi), \quad \xi = \frac{(\Delta V/2V)}{1 - (\Delta V/2V)}. \tag{8.21}$$

After the cycle, the configuration of the outer devices is initial, but the gas appears to be in non-equilibrium state. Entropy of the system can be directly estimated according to Boltzmann formulae (see section 8.2) applied to this case

$$S = k \ln W, \quad W = \frac{(2N)!}{N_1!\, N_2!},$$

so that the difference of entropy of the system in equilibrium and non-equilibrium states is

$$\Delta S = -kN\xi^2. \tag{8.22}$$

The terms of the third and higher orders are neglected here and further on.

The work ΔA which is needed to pass the system through the cycle can be calculated as a work of/on ideal gas in every of four steps of the cycle. One finds eventually that external forces have to produce extra work during the cycle

$$\Delta A = kTN\xi^2. \tag{8.23}$$

The internal energy of the system

$$U = 6NkT,$$

as internal energy of ideal gas, does not change in the process, so that the first law of thermodynamics can be written, considering every step of the process to be reversible, in the form

$$T\,\Delta S = -\Delta A, \tag{8.24}$$

and the change of entropy of the system in the process can be estimated as

$$\Delta S = -kN\xi^2. \tag{8.25}$$

Thus, as a result of this production cycle, one has a working body in a particular non-equilibrium state, which cannot be created without work of external forces and without the deliberate choice of sequence of elementary operations. Somebody possesses certain sources of energy and has an aim to create a unique non-equilibrium form of matter. To achieve the goal, the Creator sends the message in codes of elementary operations: BABA[2]. No other messages can be helpful. The information content of the deliberate message can be estimated, if one takes into account that this message is one in eight possibilities. So, the message carries the information entropy in the amount

[2] One can note that in a similar situation Maxwell's Demon had a similar aim to create a non-equilibrium state of matter. In contrast to the Creator, Demon can distinguish separate molecules and act without doing any work.

$$I = -\log_2 \frac{1}{8} = 3. \quad (8.26)$$

The information content of the message can be considered to be materialized in non-equilibrium form (complexity) of matter. The cost of materialization is the work of production equipment ΔA.

The formula for the change of entropy (8.24) does not include the amount of information I, though importance of this quantity for the creation of a non-equilibrium situation is doubtless. Contrary to the statements of some researchers, entropy and information cannot be identified. In the considered example, the variation of entropy and the amount of information defined by parities (8.25) and (8.26), are not equal, do not coincide in the form and it is not even clear how they can be compared.

8.4.2 Production of commodities

Thermodynamic production cycles, an example of which is the elementary model considered in the previous section, are designed to transform the 'wild' forms of substances into 'useful' forms (dwellings, food, clothes, machinery and so on), or, in other words, to create useful complexity.

The simple model demonstrates that, to create a proposed non-equilibrium situation of the matter, one needs purposely-designed equipment, and sources of power for deliberate operations.

8.4.2.1 Processes of production

The core of the production process is production equipment that is also known as physical production capital, measured by its value K. The production equipment is designed to be able to perform special operations, remaining (excluding wear and tear) unchanged after the cycle of transformation of some substances, the forms of which (one can assume that change of energy can be neglected) are being changed due to special work of the production equipment. A production cycle can be considered as a sequence of elementary operations $j_1, j_2, \ldots$, while the set of operations is given. The index j_l is an index of an elementary operation which is fulfilled as number l in the sequence of operations. The unique choice of indexes determines where, when and how (this is the information ΔI transferred to the system) forces are allowed to act to perform work. The total can be calculated as a sum of work at elementary operations

$$\Delta A = A_{j_1} + A_{j_2} + \cdots$$

This work is fulfilled by working people and production equipment. The inventive humans make the different appliances (production equipment), driven by outer sources of energy (wind, running water, coal, oil, ...), to execute a set of deliberate operations and work instead of humans, and the work in the production process can be written as the total of energy estimate of humans efforts L and pure work of production equipment P (substitutive work)

$$\Delta A = L + P. \tag{8.27}$$

The production equipment is instructed by humans, so that the human is eventually a creator, who is choosing a unique set of operations to provide the desired result of production process. The efforts of humans L and substitutive work P are accompanied by huge numbers of messages. The fluxes of messages about the chosen operations determine the amount of information received by the object during the production process.

The information ΔI and work ΔA appear to be the main inputs, which determine eventually a new organisation of matter that, due to production processes, acquires forms of different objects. The fluxes of energy and information bring into being a set of products that is a special complexity, which, from a thermodynamic point of view, can be characterized by reduction of entropy. The decrease in entropy of the entire environment ΔS due to processes of production can be connected with special work

$$-\Delta S = \frac{1}{T} \Delta A. \tag{8.28}$$

The change of entropy ΔS, as a result of production processes, is connected with both useful order, which is the aim of processes of production, and useless order of the environment, which is considered as waste. The useful part of entropy, which can be conveniently dubbed as $-Y$, presents a collection of products that is being studied by economists as output of the production system. For the universal quantitative estimate of products, there was elaborated a special concept—value that is usually measured with arbitrary monetary units. Neglecting contribution of non-useful waste, one can write, after the comparison of equations (8.27) and (8.28), the relation between production of value Y and total work

$$Y \sim L + P. \tag{8.29}$$

One can state that value of a set of products is determined by the sum of the efforts of the working people L and the substitution work of the production equipment P. This statement is a proper generalization of the Smith–Marx labor theory of value, which postulated that the estimate of the effort of workers L is the absolute source of all created wealth (output). It can be assumed that, if the role of production equipment is properly taken into account, the absolute measure of value ought to be the sum of the properly accounted work of people L and the work of production equipment (substitution work) P.

The more detailed discussion of economic and thermodynamic ways of description of production can be found in the author's monograph (Pokrovskii 2018).

8.4.2.2 Wealth and production factors

The efforts of working people L and the work of production equipment P leads to the creation of a set of products that is characterized by its value Y. The processes of transformation of substances and rearrangement of things (processes of production of commodities) appear possible only when fluxes of energy and information are attained to the processes. The absolute measure of the exchange value of the created products Y appears to be work $L + P$ (in power units), needed to make a thing or service. If output is measured in conventional monetary units, the relation between the output Y and production factors L and P takes a more complex form; it is demonstrated (Pokrovskii 2018) that the output Y, measured in monetary units of constant purchasing power, is completely determined by the effective utilization of production factors and can be universally represented by a function of three variables

$$Y = Y(L, P, \alpha) \quad 0 < \alpha < 1. \tag{8.30}$$

The power function is used as a specific form of the law of production of value measured in monetary units of constant purchasing power

$$Y = Y_0 \frac{L}{L_0}\left(\frac{L_0}{L}\frac{P}{P_0}\right)^{\alpha}, \quad 0 < \alpha < 1, \tag{8.31}$$

The expenditures of human efforts L and substitutive work P, which are the true sources of value, appear to be the most important production factors[3]; the quantity α, called the technological index, shows the effectiveness of using factors of production and is a characteristic of the production system. The values L, P and α are, generally speaking, functions of time and are defined by a system of equations (Pokrovskii 2018, chapter 5, equations (5.36)). The development of the production system is determined, eventually, by the possibilities of attracting workers and additional external energy to the production processes, and the possible increase in human effort and substitute work is always the focus of human society. The details of the derivation and discussion, also as comparison with real data for the US and Russia, can be found in the author's monograph (Pokrovskii 2018).

To give an idea of the scale of human productive activity, we write out a direct estimate of the value of the world's social wealth (Shorrocks *et al* 2010, table 2.4) for 2010; it is about 2×10^{14} US dollars, and about half of this amount is the real (not financial) wealth. This figure presents wealth of all kinds (material and non-material, personal and industrial) and indicates a huge accumulation of artificial objects, among which people live.

8.4.2.3 About absolute measure of value

From relation (8.29) it follows that the natural unit for measuring of created value Y is energy, and, consequently, universal energy measure of value can be introduced. Certainly, it does not mean that value is equivalent to energy; as we saw, value is a close relative to entropy with the 'minus' sign. From the thermodynamic point of view, value of a set of products can be interpreted as complexity of this set created by

[3] It should be noted that the use of huge amounts of energy in the production of goods is impossible without the presence of *production equipment*, which is measured by its value K. This quantity, also called physical production capital, should be considered as a factor (passive, in contrast to the factors L and P) of production. It has been observed that the use of production equipment (in other words, the attraction of production capital) increases the productivity of labor, which gave rise to the myth of the productive power of capital.

efforts of people and substitutive work of production equipment. In economic practice, the exchange value of created products is estimated by arbitrary monetary units, at best, monetary units of constant purchasing power, and the question arises about the relation of monetary units with absolute energy units.

In the Smith–Marx theory of value, it was postulated that estimate of the efforts of workers L is the ultimate source of all created wealth (products), an absolute measure of value. When one accounts for the effect of substitution of workers' efforts with the true work of the production equipment, one could expect that the total amount of work on the production of value, shown by equation (8.28), that is, the sum of properly accounted work of humans L and work of production equipment (substitutive work) P, both measured in power units, could be an absolute measure of value. An assessment of works (energy) necessary for creation of a product valued by one monetary unit, or, in other words, 'the power content of a monetary unit', represents a universal, independent from place and time, energy measure of value.

The work required to create a product that represents a single monetary unit, or in other words 'energy content of the monetary unit', can be found as the ratio of the work (8.27) to the output Y, measured in monetary units of constant purchasing power. This ratio depends on the factors of production L and P and changes over time; evaluation of the 'energy content' of the international dollar, according to estimates of global gross domestic product since the beginning of our era to the present day shows (Pokrovskii 2018, chapter 12) that in ancient times, when handwork prevailed over the machine, $L \gg P$, the energy content of the monetary unit had a tendency to decrease, but in the industrial era, when work outweighed the efforts of the workers, $P \gg L$, the energy content increases. The estimated current values display (Beaudreau and Pokrovskii 2010) that the average value of the 'energy content' of dollar year 1996 in the period 1960–2000 is 1.4×10^5 J, while the average value of the 'energy content' of the ruble 2000 in the same time interval (1960–2000) is 0.1×10^5 J. These values coincide within the accuracy of economic measurements, taking into account purchasing power parity of the dollar and ruble estimates, which confirms that the energy measure of value appears to be the universal measure. These assessments ultimately confirm the theory of social production based on the generalized labor theory of value (Pokrovskii 2018).

References

Beaudreau B C and Pokrovskii V N 2010 On the energy content of a money unit *Physica A: Stat. Mech. Appl.* **389** 2597–606

Boltzmann L 1877 Über die Beziehung zwischen dem Zweiten Hauptsatze der mechanischen Wärmetheorie und der Wahrscheinlichkeitsrechnung resp. den Sätzen über das Wärmegleichgewicht *Sitzungsber. Kaiserl. Akad. Wiss. Wien Math- Naturwiss. Kl.* **76** 373–435 Also:

Boltzmann L 1964 *Lectures on Gas Theory, English Translation* (Berkeley: University of California)

Brillouin L 1956 *Science and Information Theory* (New York: Academic)

Kondepudi D and Prigogine I 1999 *Modern Thermodynamics: From Heat Engines to Dissipative Structures* (Chichester: Wiley)

Landau L D and Lifshitz E M 1986 *Statistical Physics* vol 1 3rd edn (Oxford: Pergamon)

Maxwell J 1871 *Theory of Heat* (London: Longman Green) Also:
Maxwell J 1970 *Theory of Heat* (Westport, CT: Greenwood)

Pokrovskii V N 2018 *Econodynamics: The Theory of Social Production* 3rd edn (Dordrecht: Springer)

Shannon C E 1948 A mathematical theory of communication *Bell Syst. Tech. J.* **27** 379–423 and 623–56

Shorrocks A, Davies J B and Lluberas R 2010 *Global Wealth Databook 2010* (Zurich: Credit Suise Group AG)

Simon H A 1996 *The Sciences of the Artificial* 3rd edn (Cambridge, MA: MIT Press)

Sommerfeld A 1952 *Thermodynamik und Statistik—Vorlesungen uber theoretische Physik Band 5* ed v F Bopp and J Meixner (Wiesbaden: Diederichsche Verlagsbuchhandlung) Translation: Sommerfeld A 1964 *Thermodynamics and Statistical Mechanics—Lectures on Theoretical Physics* vol. V ed F Bopp and J Meixner (New York: Academic) translated by J Kestin

Tribus M and McIrvine E C 1971 Energy and Information *Sci. Am.* **225** 179–88

Trincher K S 1965 *Biologiya i informatsiya: Elementy biologicheskoy termodinamiki [Biology and Information: Elements of Biological Thermodynamics]* 2nd edn (Moscow: Nauka) revised and extended

Chapter 9

Problems of thermodynamics of living organisms

A living biological organism intensively exchanges heat and substances with the environment and appears to be an example of an open thermodynamic system; it exists, naturally, in a non-equilibrium state and comes to the equilibrium state only after death. The organism contains a huge number of chemical compounds and represents a complex dynamic structure, emerging as a result of some deliberate chemical reactions. The brusselator described in chapter 6 can be considered as one of the elementary models, which reflects the most essential features of the biological organism: exchange of substances with the environment and existence of the stable spatial internal structure within the limited volume, where chemical reactions and processes of the substance translocation are occurring. The thermodynamic theory allows formulation of the equation of growth for the living biological organism.

9.1 The living organism as a thermodynamic system

It is apparent that there are no doubts that the principles of non-equilibrium thermodynamics can be applied to description of functioning and development of the living organism (Schrödinger 1944, Prigogine 1961, Trincher 1965, Zotin 1972 and many others); nevertheless, the thermodynamic system that can be attributed to the object, has some special features. The main peculiarity of the living object, which separates it from other thermodynamic systems, is activity of the living object; there is a special inherent program that governs the functioning of the system via transformations of chemicals. The content of the fluxes between an organism and the environment is a choice of the organism. Though the thermodynamic laws are applicable, the behaviour of the system is prescribed by the inherent program, and it brings some peculiarities into the description of the organism in terms of thermodynamics.

doi:10.1088/978-0-7503-3451-8ch9

9.1.1 Organism as an ensemble of chemical reactions

The biological organism contains a huge number of chemical compounds, which can be numerated by index α. We assume that the all substances are participating in chemical reactions, so that, in fact, the living biological organism can be considered as an ensemble of coupled chemical reactions, and, to represent thermodynamic functions for the biological organism as for the ensemble of chemical reactions, one can refer to the results of chapter 6. At the assumed constant temperature and pressure, according to the results of section 6.2, the expression for the changes of Gibbs free energy and entropy of the system ought to be written as

$$dG = -\sum_{j} A_j \, \Delta\xi_j + \sum_{\alpha} \mu_\alpha \, \Delta n_\alpha, \tag{9.1}$$

$$T \, dS = \Delta Q - \sum_{j} \Xi_j \, \Delta\xi_j + \sum_{\alpha} \eta_\alpha \, \Delta n_\alpha, \tag{9.2}$$

These equations define the thermodynamic functions G and S for the living organism at permanent exchange of substances and energy with the environment. The first term on the right-hand side of equation (9.2) presents a stream of heat energy into the system; it can be positive (the stream into the system) or negative (the stream out of the system). The last terms in equations (9.1) and (9.2) depict streams of chemical energy incoming or outgoing with the stream of particles of substances Δn_α through the borders of the system. There is no need to account for the change of intermediate substances within the system, so that the index of environment in the symbols for the change of the numbers of molecules n_α in the last terms of equations (9.1) and (9.2) can be omitted. The quantities A_j and Ξ_j in equations (9.1) and (9.2) are, correspondingly, the affinity of reaction with label j, defined by equation (6.12), and the thermodynamic force, defined by equation (6.16) in chapter 6. The quantities μ_α and η_α denote the chemical potential and the partial entropy carried by the substance labeled α. According to equation (2.27), the sum of chemical potential and partial entropy results is specific enthalpy of the substance, $h_\alpha = \mu_\alpha + \eta_\alpha$. Internal variables ξ_i in the first term of equations (9.1) and in the middle term of equation (9.2) (in the right parts of the equations) are measures of incompleteness of chemical reactions, that is, the measures of how much the system with chemical reactions is far from equilibrium. Generally, the set of internal variables ξ_j for an alive organism should include not only measures of incompleteness of all chemical reactions occurring in the system, but also gradients of temperature, a difference of concentration of substances and other characteristics.

Values of thermodynamic functions (9.1) and (9.2) increase or decrease depending on variations of internal variables ξ_i which appear at external influences ($\Delta\xi_j > 0$) and disappear (relax according to internal laws) ($\Delta\xi_j < 0$), being dismissed. For small deviations of internal variables from their equilibrium values, which are considered as equal to zero, the local law of disappearing can be written as a relaxation equation for each internal variable

$$\frac{d\xi_i}{dt} = -\frac{1}{\tau_i}\xi_i, \quad i = 1, 2, \ldots ,$$

where $\tau_i = \tau_i(T, x_1, x_2, \ldots, x_n)$ is a relaxation time of the corresponding internal variable. The dynamics of internal variables for arbitrary deviations from equilibrium was formulated in chapter 3. The examples of oscillating behaviour of internal variables can be found in section 6.3.

When internal variables are emerging, the value of Gibbs free energy increases and value of entropy decreases, and, at the same time, because of internal processes of a relaxation, value of Gibbs free energy decreases and value of entropy increases. The processes of creation and relaxation of internal variables are balanced in a non-equilibrium stationary state of the system, when $dG = 0$ and $dS = 0$. In non-stationary situations, one of the processes could prevail, and one can observe creation or distraction of the internal variables.

9.1.2 Hierarchy of internal variables

The structure of a living organism appears to us as a set of internal variables ξ_j, which are a measure of the incompleteness of chemical reactions in a living organism. Under the influence of flows of substances from the environment, the internal variables ξ_i are excited; the system comes to an unbalanced state. These processes are opposed by the tendency of internal variables to equilibrium values, which is the driving force of all internal processes in the system. The rate of disappearing of the internal variables varies; times of relaxation of some of internal variables are very small as compared with the lifespan of organism, but other internal variables exist during the total lifetime and even more. The differences of internal variables according to their rates of approaching the equilibrium state determines their functional distinctions in the processes of the growth and development of the organism[1]. For further discussion, we shall divide, following the work (Zotin and Pokrovskii 2018), the set of internal variables of the system into three groups, according to values of relaxation times; later, the different functional roles of separate groups in the processes of development and growth are described.

The 'eternal' internal variables

First of all, it is possible to separate a group of internal variables, which originates from the complexity of the first cell that gives the beginning of the organism. The material carriers of complexity are the molecules of a deoxyribonucleic acid packed in the nuclei of the cells. The copies of internal variables (complexity) of the initial set are present in each cell and remain unchanged (at some idealization) up to the death of an organism. The number of copies of these variables increases at the proliferation of cells, and during the realization of the program of development. These variables comprise the foundation of a control mechanism that supervises all

[1] Consideration of a hierarchy of relaxation processes is useful in many applications of non-equilibrium thermodynamics. For example, Gujrati (2018) demonstrated recently that the division of internal variables according to their relaxation times is very helpful for description of the phenomena of solidification of glasses.

processes in the organism during its existence and growth. The organism is developing under an installed plan received by right of succession, whereas some internal variables are suppressed or expressed in the multiplying cells. A role of internal variables of the first group is the conservation of hereditary instructions and control over the construction of an organism. The complexity of the control device cannot be less than the complexity of the object that is controlled, so that the main carrier of complexity is the internal variables of the first group.

The internal variables of structure

The second group of internal variables describes the emerged structure of the organism in processes of morphogenesis and differentiation, when the interoperability of the parts of an organism is materializing, which provides such processes that cannot occur in a random mixture of chemical reagents. The number and values of internal variables of this group increase at growth and development of the organism. Eventually, the set of variables of the second groups determines some metastable structure (organs and tissues) of the living organism. These internal variables have the times of relaxation comparable with the lifespan of the organism, so that they do not contribute into the current dissipation of energy (production of entropy).

The fast-relaxing internal variables

The set of internal variables of the third group emerges in the course of the above-described processes of multiplication of original complexity and construction of the structure of the organism. The processes are based on chemical transformations of substances, occur in non-equilibrium situations and are accompanied by excitation of internal variables and their fast relaxation. The set of internal variables of this group includes, apart from degrees of completeness of fast chemical reactions occurring in the organism, the variables locating gradients of temperature, differences in concentration of substances and so on. Internal variables of the third group accompany all existing processes anywhere they were. These internal variables have very small times of relaxation (in comparison with existence of the organism) and determine the basic contribution to the internal production of entropy. The dissipated energy leaves the system with outgoing fluxes of substances and coming out directly in the form of heat. Specific energy dissipation usually decreases with the growth of the body, but during periods of intensive growth (early development, regeneration, malignant tumors), energy dissipation increases (Zotin and Zotin 1997).

9.1.3 Complexity and entropy

The classification of internal variables allows us to record a schematical expression for entropy. The living organism can be perceived as the metastable structure formed by the set of internal variables of the second group $\boldsymbol{\xi}^{(2)}$; it is possible to consider these slowly-varying variables as parameters of a metastable state of the system. The deviations of the system from this state are described with vastly-varying variables of the third group, which, as it is possible to judge according to the amount of thermal emission (dissipation), can be considered as the small quantities. Referring to expressions (4.41) and (6.24) as samples, one can present entropy S of the organism in the elementary approximation as a square-law function of internal variables of the third group

$$S(\boldsymbol{\xi}^{(1)}, \boldsymbol{\xi}^{(2)}, \boldsymbol{\xi}^{(3)}) = S_0(\boldsymbol{\xi}^{(1)}, \boldsymbol{\xi}^{(2)}) - \frac{1}{2} S_{ij}(\boldsymbol{\xi}^{(2)})\, \xi_i^{(3)} \xi_j^{(3)}. \tag{9.3}$$

The quantity S_0 is a decreasing function of variables of the first group $\boldsymbol{\xi}^{(1)}$ (carriers of instructions for construction and functioning of an organism) and the second group $\boldsymbol{\xi}^{(2)}$ (the description of metastable structures of the organism). The matrix S_{ij} is positive-definite and essentially depends on internal variables of the second group (the metastable structure of the system) that cannot be considered as small quantities.

In line with entropy, concept of *complexity* that is discussed in section 8.1.1, can be used; the change of complexity can be connected with the change of entropy of the system by equation (8.1). The complexity can be considered as a characteristic of the set of internal variables ξ_i; the internal variables themselves are sometimes called *variables of complexity* or *variables of order*. Though it cannot yet be defined how complexity is related to the number and connectedness of internal variables, one can assert that the greater the number of internal variables, their values and mutual connections, the greater the complexity. Equation (9.3) describes that increase or decrease in number and values of internal variables provokes decrease or increase in entropy.

The concept of complexity has the same sense as notion of *bound information* (Brillouin 1956) in application to complex systems. The researchers evaluate entropy of the organism by calculating the 'information' of the system[2], but the assessments of such a kind are influenced with subjective understanding what sense has concept of *the information* in application to the thermodynamic system (see section 8.3).

9.2 The growth and development of the living organism

The thermodynamic approach gives a clue to the problem of description of the growth and development of a biological organism from an impregnated embryo, to a completely developed organism (the theory of ontogenesis). We shall discuss the

[2] Entropy of the architectural order, for example, has been estimated by Blumenfeld (1981). His assessments show, that the contribution of the architectural component is insignificant, also he has come to the conclusion, that 'all conversations about anti-entropy tendencies of biological evolution, about outstanding orderliness of the living matter are based on misunderstanding. According to thermodynamic criteria, any biological system is ordered no more, than a piece of rock of the same weight.' (Blumenfeld 1981, p 27). Such a paradoxical conclusion is based, certainly, on misunderstanding: Blumenfeld has not considered the contribution of a functional component to entropy of the organism. However, after the assessments of the architectural entropy, some reasoning of the author (Blumenfeld 1981, p 27) about biological orderliness follows: 'it is difficult to get rid of consciousness, that the deceit is covered somewhere: it is clear visible, without any calculations, in fact, the amazing orderliness of biological structures and processes!'. The 'orderliness of structures and processes' is just what has been neglected by Blumenfeld estimating entropy of the biological organism.

processes of growth and development of the living organism, considering it as an open thermodynamic system of undefined volume, but with the certain mass M and temperature T. The permanent exchange of substances and energy with the environment is needed for the biological organism to grow. The growth occurs under control of an inherent program and stops when the organism takes the prescribed-by-the-program form.

9.2.1 The growth equation

Equations (9.1) and (9.2) connect changes of Gibbs free energy G and entropy S of a living organism with fluxes of substances and heat energy and transformation of internal structure. The rate of growth of mass M in the thermodynamic terms can be connected to heat and amount of substances that are measured in laboratories. The heat ΔQ can be measured directly, but the energy carried with the every unit of substances Δn_α is estimated in laboratories as specific enthalpy h. This amount of energy contributes to Gibbs free energy G and entropy S as chemical potential μ and specific entropy η.

To correspond with the measured quantities, one has to consider enthalpy of the system $H = G + TS$. Values of the extensive thermodynamic functions increase with the growth of the organism, and, to exclude the growth connected with the increase of weight of the entire organism M, it is convenient to consider specific thermodynamic functions, which are, in particular, H/M, which is the characteristic of the entire organism $H = G + TS$, divided by its weight M. On the basis of fundamental expression for variation of Gibbs free energy and entropy (9.1) and (9.2), the equation for the growth rate of the living organism can be written as

$$\frac{1}{M}\frac{\mathrm{d}M}{\mathrm{d}t} + \frac{M}{H}\frac{\mathrm{d}}{\mathrm{d}t}\left(\frac{H}{M}\right) = \frac{1}{H}\left(\frac{\Delta Q}{\Delta t} - \sum_j (\Xi_j + A_j)\frac{\mathrm{d}\xi_j}{\mathrm{d}t} + \sum_\alpha h_\alpha \frac{\Delta n_\alpha}{\Delta t}\right). \tag{9.4}$$

It is marked with the symbol Δ that the corresponding quantities are objects of direct empirical observation.

The law of conservation of mass also determines the equation of growth

$$\frac{\mathrm{d}M}{\mathrm{d}t} = \sum_\alpha m_\alpha \frac{\Delta n_\alpha}{\Delta t}, \tag{9.5}$$

where m_α is mass of a particle of substance labeled α. Unfortunately, this form of the growth equation (9.5) is useless practically, as there are a lot of the processes occurring in the living system, and it is extremely difficult to measure the substances coming in and out the organism (Zotin 1972).

The balance equations (9.4) and (9.5) provide a base for thermodynamic interpretation of the phenomena of existence, growth and ageing of a biological organism.

It is convenient to introduce a special symbol for the growth rate of specific function

$$\frac{M^2}{H}\frac{\mathrm{d}}{\mathrm{d}t}\left(\frac{H}{M}\right) = -\sigma, \qquad \sigma \geqslant 0. \tag{9.6}$$

General consideration does not determine the value and behavior of σ, but one can rely on the universal mechanism of growth of a biological organism by cell division. The universal growth mechanism assumes that an organism's cell divides into two cells, while each new cell contains a complete program that controls growth and development. New cells can acquire new functions (differentiation) and form new organs (morphogenesis). An increase in the number of cells corresponds to an increase in the size and weight of the body. The growth and development of a living organism is associated with the emergence of a new complexity and an increase in body mass. We can think that the complexity associated with the internal variables of the first group prevails over the complexity associated with the variables of the second group, which leads to the constancy of the specific thermodynamic functions of the system, which may be due to the peculiarities of the mechanism of growth of a living organism by cell division.

Then, given the definition (9.6), *the growth equation* (9.4), can be written in the form

$$\frac{\mathrm{d}M}{\mathrm{d}t} = \sigma + \frac{M}{H}\left(\Delta Q - \sum_j (\Xi_j + A_j)\,\Delta\xi_j + \sum_\alpha h_\alpha\,\Delta n_\alpha\right). \tag{9.7}$$

This equation is a reformulation (in the term of enthalpy) of the growth equation that was derived earlier (Zotin and Pokrovskii 2018).

The growth is occurring according to the inherent program, until the organism appears to be adult, when $\Delta M = 0$; then the organism ought to be considered as a stationary system[3], in which all processes appear to be balanced

$$\begin{aligned} -\sum_j A_j^{(3)}\,\Delta\xi_j^{(3)} + \sum_\alpha \mu_\alpha\,\Delta n_\alpha &= 0, \\ \Delta Q - \sum_i \Xi_i^{(3)}\,\Delta\xi_i^{(3)} + \sum_\alpha \eta_\alpha\,\Delta n_\alpha &= 0. \end{aligned} \tag{9.8}$$

We can assume that the structure of the system appears to be unchangeable in the stationary state, so that the internal variables of the first and second group do not contribute to the balance, due to their definitions.

The drive power of the development is the tendency of the organism to absorb enough of the substances containing building materials and energy. The fluxes of substances are governed by the inherent program (instruction) of the organism. The incoming streams of substances leads to emerging of internal variables of the second and third group. The relaxation of variables of the third group defines dissipation of energy that via fluxes of substances and in thermal form is completely deleted from

[3] This statement is not valid for all organisms: the growth of some reptiles (crocodiles, for example) does not stop.

the organism[4]. For existence and development of the organism, except for the flux of energy in the form of fluxes of substances, an exchange of heat with the environment is needed. It is possible to say about the flux of heat from the organism, using the words of (Schrödinger 1944), that from the thermodynamic point of view the organism 'eats negative entropy'.

9.2.2 The testing of the growth equation

Extensive studies of the growth and development of various organisms indicate a variety of types of growth and development of an individual organism (Zotin 1972). The equation of growth (9.7) in its general form is applicable for any living organism. However, due to peculiarities of growth, the growth rate of specific function (9.6) ought to be defined for chosen species. Leaving aside investigation of the growth of separate cells and microorganisms (von Stockar and Liu 1999, von Stockar *et al* 2006), we turn to many-cell organisms that have a special mechanism of the growth by division of cells. To test equation (9.7), Zotin and Pokrovskii (2018) have referred to direct observation of the growth of animals in situations, when it is possible to control an exchange of heat and incoming flux of energy into the organism through the fluxes of substances.

The methods of estimation of energy fluxes

The thermodynamic equation of growth (9.7) asserts that the difference between the quantities of incoming-in-chemical-form energy and outgoing heat is used for the increase in the mass of the organism. The statement is valid for cases when researchers try to exclude fluctuations and random factors, so that the statement concerns to the basic exchange between the organism and the environment. Methods of assessment of the incoming and outgoing flows of energy, presented in the right-hand part of equation (9.7), are well developed (Zotin 1972). The flux of heat energy from the living system ΔQ (the first term in the right-hand part of equation (9.7)) is measured by *the method of direct calorimetry*, that is, by direct measurement of the heat production of the organism by means of a calorimeter. The amount of thermal energy coincides with the dissipation of energy that appears at the relaxation of internal variables of the third group with small times of relaxation. The amount of the heat flux is evidence of the presence of quickly relaxing internal variables.

The incoming energy comes into an organism of animals with fluxes of substances in the form of energy of chemical links of organic molecules; the greatest amount of energy goes together with food. The fluxes of energy going in and out with the substances cannot be accounted for directly; instead of it, the amount of the

[4] It is known (Zotin and Zotin 1997) that specific dissipation, that is the amount of thermal emission (dissipation) divided by the mass M of the organism, decreases, as a rule, with growth of an organism. The approach to steady state is accompanied by decrease in internal production of entropy, but let us note the difference: if the inherent program is absent the decrease of dissipation is connected with decrease in the values of internal variables, but in the opposite case, the decrease in production of entropy is connected with the restructuring of the set of internal variables, whereas the number of internal variables with large relaxation times increases.

coming-with-food energy is being estimated as a quantity of adenosine triphosphate (ATP) that is synthesised within the organism. The synthesis can be realized in the two basic processes: the oxidizing phosphorylation and glycolysis. The energy of molecules of ATP is used basically for the synthesis of molecules, which are included in the organism, also as for other processes requiring expenses of energy (muscle contraction, migration of cells, conduction of nervous impulses, etc). Regarding the above-described classification of internal variables, the result of the release of energy of molecules of ATP is the emergence of internal variables of the second group, which is accompanied by excitation of internal variables of the third group.

Moreover, the quantity of ATP cannot be detected directly, the speed of synthesis of adenosine triphosphate (ATP) is estimated indirectly (in a unit of time). If one assumes that the quantity of ATP, received in the process of glycolysis, can be neglected in comparison with the process of oxidizing phosphorylation, then one judges about the speed of synthesis ATP according to the rate of oxygen consumption or the rate of allocation of carbonic gas (Zotin 1972). It is shown that quantity of the consumed oxygen is proportional to the energy coming into the organism, whereas the factor of proportionality (so-called oxi-caloric coefficient) is approximately equal to quantity of 4.8 cal per one ml of the absorbed oxygen. This procedure has received the name *the method of indirect calorimetry*.

The accuracy of the measurements of fluxes of energy by the described methods are not high; it has been discovered, for example, that during an essential calm period of ontogenesis of organisms, the values of incoming and outgoing energy fluxes practically coincide (Zotin 1972). However, there is mass growth in such situations, and, according to equation (9.7), the difference between incoming and outgoing flows of energy must exist. One can assume that the difference appears to be so small, that, apparently, it cannot be detected with the used methods. Apart from that, when the method of indirect calorimetry is used for the assessment of incoming stream of energy, the contribution of processes of glycolysis, which are also processes of synthesis of ATP, are neglected in comparison with the process of oxidizing phosphorylation. However, the contribution of the processes of glycolysis can appear comparatively essential in this phase of calm growth.

Estimation of entropy of living organisms

Due to difficulties of the measurements, works that are aimed at the comparison of the results of two methods of measurement of the energy exchange are rare. The authentic distinction between data of the indirect (according to the rate of oxygen consumption) and the direct (according to the rate of heat production) calorimetry in situations of the constant environment is observed in cases of the initialization of such processes, as differentiation, morphogenesis etc (Grudnickij and Nikol'skaja 1977, Zotin and Zotina 1993, Kleymenov 1996). Estimating the external fluxes, Kleymenov (1996) has detected, that 'for all the investigated objects in early ontogenesis, irrespective of character of growth and taxon accessory, the intensity of the energy exchange, measured by a method of indirect calorimetry, steadily exceeds data provided by the direct calorimetry'. Kleymenov informs us also that the difference between measurements by two methods reliably correlates with specific growth rate of the animals, and there are no authentic distinctions in value of factor of proportionality between a difference of readings of indirect and direct calorimetry and the growth rate of various organisms. Comparing the reported average value of coefficient of proportionality with that in equation (9.7), $(G + ST)/M = 770$ kJ g^{-1}, allows us to estimate the top value of the specific entropy of the organism, at assessment of temperature in energy units, as $S/M \approx 2 \times 10^{26}$ g^{-1} or, at assessment of entropy in energy units, as $S/M \approx 2500$ J g^{-1}. Estimation of the top absolute value of specific entropy of different organisms shows (Zotin and Pokrovskii 2018) that this quantity decreases in an evolutionary sequence of organisms: yeast → insects → reptiles → birds. Complexity of the living organism should be connected with the value of specific entropy S/M: the less the specific entropy, the greater the structural complexity of the organism.

The analysis of the available data shows also (Zotin and Pokrovskii 2018) that value of the quantity σ in equation (9.7) appears to be about zero. It implies that specific thermodynamic functions, in particular, specific entropy of the organism S/M, do not change during its growth and development, which confirm the universal mechanism of growth by dividing cells, while each new cell contains the total program that controls the growth and development. The new cells can acquire new functions (differentiation) and comprise new organs (morphogenesis). It can be described as multiplication of internal variables of the first group and emerging of structural variables of the second group. The growing complexity could lead to reduction of specific entropy of the system, but, entropy of the whole organism, at the same time, increases with the increase of weight M. Under the balance of incoming and outgoing energy, the specific entropy S/M, can only decrease. In situations of growth, the complexity of organism increases, and, so as the greater the complexity, the less entropy, one can assume that the specific entropy does not increase, and it gives that value of the quantity σ appears to be about zero.

9.3 Some peculiarities of the living organism

Equations (9.1) and (9.2) define the thermodynamic functions for the entire living organism, which, however, can be also considered as a collection of separate organs within the living organism. Therefore, thermodynamic functions for the whole system can be presented as the sum of the thermodynamic functions of corresponding subsystems (Gladyshev 2017), whereas each subsystem ought to be described as the open one. To describe particular processes, many researchers (for example Belintzev 1991, Turing 1952) have introduced specific variables (which belong to the second group of variables in our classification) within the organism to investigate possible mechanisms of increase of number of cells, spatial differentiation of forms, structuration and so on. The modern review of some results can be found, for example, in the book by Romanovskii with co-authors (2003). Due to the huge complexity of the system, we are still very far from a complete description of the living organism as a dynamic system; we cannot even inventory all internal variables. Nevertheless, the researches succeeded in describing certain characteristic features.

9.3.1 The control and regulate system

From a thermodynamic point of view, the biological organism is a dynamic system; the structure and functionality of the system are described with the set of internal variables ξ_j, and, moreover, in our imagination, the identity of the living organism is connected with the internal variables of the second group (see section 9.1.2). The set of these variables $\xi_j^{(2)}$ describes the metastable structure (organs, tissue, bones, etc) and functionality of the organism. The properties of the *external environment*, in which the organism exists, are widely varied, whereas the parameters of the internal state of the organism are being kept within certain preset limits. A living organism can be situated, for example, in the environment of various temperatures, but temperature of the body of the organism remains at about a value that is characteristic for the organism. The internal state of the living organism is essentially independent of the external environment. The stability of the internal state is a condition of existence of the organism, and such a phenomenon has the name of *homeostasis* (Cannon 1932). To maintain the internal environment of the organism in the stable state, the organism should have a control and regulate system that can be schematically described in thermodynamic terms.

We consider that Gibbs free energy and entropy of the organism is constant and the system is stable. Assuming the condition (9.8) is valid in an unperturbed state, we can find that deviations of the internal variable $\xi_j^{(2)}$ and fluxes from the stationary values should obey the relation

$$\delta Q - \sum_i \left(\Xi_i^{(2)} + A_j^{(2)}\right) \delta\xi_i^{(2)} + \sum_\alpha h_\alpha \, \delta n_\alpha = 0. \tag{9.9}$$

A change of one of the variables from the list of the internal variables $\xi_j^{(2)}$ or fluxes of heat and matter is a signal for the internal control and regulate mechanism to change values of other variables (remember that the content of the fluxes between organism and the environment is a choice of the organism). At the rise in outer temperature, for example, the flux of heat Q changes, the control mechanism receives the signal and, for temperature of the organism not to be increased, the regulate mechanism commands (that is connected with alternation of the internal variable $\xi_j^{(2)}$) an increase of perspiration (the change of the third term in equation (9.9)).

There are a great many homeostatic mechanisms, encompassing many aspects of human physiology, that control the composition of the extracellular fluid (that is known as the internal environment) with regard to the temperature and the concentrations of different substances. The control and regulate system includes the nervous system of the organism, which operates as an integrative system, connecting in a single whole sensitivity, impellent activity and the work of other regulate systems (endocrine and immune).

The description of the control and regulate system of the living organism reminds us of the famous Le Chatelier's principle, which states that the disturbed system will adjust itself in such a way that the effect of the change will be reduced or moderated. Le Chatelier's principle is valid for systems in thermodynamic equilibrium; is it possible to generalize the principle for the system in non-equilibrium steady state?

9.3.2 Mechanism of mobility

To maintain its own existence, a biological organism has to fulfill mechanical work, moving its own body, or moving foreign bodies. The movement is one of functions of the organism, and work is being done due to chemical energy, incoming into the organism with various substances. For transformation of chemical energy into mechanical energy, 'mechanochemical devices' are incorporated into biological organisms (Bisio *et al* 2003). A set of elementary 'mechanochemical devices' are acting within a biological organism doing work, but for consideration, we shall separate one 'mechanochemical device' which can be presented in the simplest way as a fibre immersed in a mixture of some reacting substances. The fibre has the length x that is capable of reducing and extending at variation of the degree of completeness of chemical reaction ξ. One end of the fibre is fixed, and the second is mobile, so that the fibre can produce work. The 'mechanochemical device' operates cyclically in isothermal conditions, and the characteristic of the device is work accomplished for one cycle.

Free energy of the 'mechanochemical device' depends on length of the fibre x and degrees of completeness of chemical reaction ξ; parities (2.20) and (6.12) allow us to record the expression for the differential of free energy in the form

$$dG = X(x,\, \xi)\, \mathrm{d}x - A(x,\, \xi)\, \mathrm{d}\xi, \tag{9.10}$$

where A represents affinity of the chemical reaction. The condition of functioning of the 'mechanochemical device' appears to be the fact that the force X depends on the degree of completeness of chemical reaction ξ and affinity of chemical reaction A depends on length of a fibre x, whereas

$$\frac{\partial X}{\partial \xi} = -\frac{\partial A}{\partial x}.$$

Let us pay attention to the fact that the parity (9.10) describes a non-equilibrium situation.

Let us imagine a cycle starting from the state $(x^{(1)},\, \xi^{(1)})$ of the system, so that the force and affinity have values $X^{(1)}$, $A^{(1)}$. It is possible to imagine, that such a solution is chosen, that, at addition of some substance into the system, the value of a degree of completeness of the chemical reaction decreases to $\xi^{(2)}$, which causes an increase of length up to $x^{(2)}$. The chemical reaction proceeds further, and the degree of completeness increases to former value $\xi^{(1)}$, then the length is reduced to former value $x^{(1)}$. The 'mechanochemical devise' has returned to the initial state and the cycle repeats.

The work made by the system in one ideal cycle, is calculated by the integration of expression (9.10)

$$\Delta W = -\oint X \mathrm{d}x = \oint A \mathrm{d}\xi \approx (A^{(2)} - A^{(1)})\Delta\xi. \tag{9.11}$$

Work is completely defined through characteristics of the reacting mixture. We shall notice, that the degree of completeness of chemical reaction $\Delta\xi$ can be expressed through number of molecules of the substances participating in the reaction (see formula (6.5)), affinity of reaction A is the weighed sum of chemical potentials of all substances participating in the reaction (see formula (6.12)), so that parity (9.10) can be specified for a particular case.

The existence of a set of 'mechanochemical devices' allows the organism to produce continuously controllable work W due to the delivery of the chemical energy taken by an organism from an environment.

9.4 Evolution of living matter

The theory of biological evolution aims to find the beginning of life and to describe the existence of a variety of living organisms on the Earth (Volkenshtein 1984). It is possible to think that the time has passed since there was the belief that development of complexity is incompatible with the law of increase of entropy; evolution does not contradict the principles of non-equilibrium thermodynamics. Under abundance of energy due to *the mechanism of delayed relaxation*, the transition to a more organized state of matter is possible as a natural process (see section 6.4). After the first steps of self-organisation of matter, the mechanism of retaining and reproducing complexity had emerged, which allowed expansion of living matter over the Earth. The thermodynamic laws and the energy principle of evolution operate and will operate progress of living matter on the Earth.

9.4.1 Ontogenesis and phylogenesis

It was demonstrated in this chapter that the growth and development of a separate organism (*ontogenesis*) can be understood from the thermodynamic point of view, but to describe biological evolution, which is the process of occurrence of new and more complex types (*phylogenesis*), we should consider a separate individual as a representative of a population of the certain type that should be the object of investigation. Each individual of a population in each cell of its organism has a unique program of functioning and development, which was received from its parents. The program assumes transfer of the combined complexity of two organisms to a new organism that appears as a member of the population. As a whole, the programs of all organisms form a genofond of the population. The genofond of a population defines the biological type. The genofond of a population is also the thermodynamic complexity that is passed from generation to generation.

The congenital program is stable, and it is possible to consider the program to be constant during all time of the development and life of the organism. But changes occurs; as a result of errors in copying or under action of thermal motion, small variations (mutation) appear, which are also passed by right of succession. These modifications have been collecting and, eventually, lead to essential variation of organisms. According to the *energy principle of evolution*, described in chapter 11 of the monograph (Pokrovskii 2018), those populations, which can utilize the greater amount of energy from their environment have an advantage for survival; thus, a part of species perishes, not having left posterity, and some manage to survive—a population of a new type of organisms emerges. In this way there are various types of living organisms with various programs of development on the Earth. This process has been continuing to our days.

9.4.2 Where is the beginning?

The basic debate has arisen in discussion of a question: how could it happen that the initial assembly of accidental substances had evolved to the modern high-organized state of the biosphere (Gladyshev 2017, Popovich 2018). This process does not contradict principles of non-equilibrium thermodynamics; in the early days of the Earth there was an abundance of potentially available energy sources, coming eventually from the Sun, and the transition to a more organized state of the matter appeared to be possible as a natural process (see section 6.4). Of course, that was the business of chance: among the chaotically arising substances, there are some that could devour a lot of energy. In work by Michaelian (2011), for example, it is specified, that molecules of ribonucleic and deoxyribonucleic acids (RNA and DNA, the macromolecules, providing storage, transfer from generation to generation and realization of the genetic program of progress and functioning of an organism) are the most effective in comparison with all known molecules to absorb intensive ultraviolet light which got through a dense early atmosphere of the Earth. Karo Michaelian has presented the possible script of the first steps of life, according to which replication and metabolism (connected from the beginning) are initiated by streams of energy. The initial complexity emerged, and later the mechanism of storage and transfer of complexity was developed.

References

Belintsev B N 1991 *Fizicheskie osnovy biologicheskogo formoobrazovaniya, Pod red. M.V. Vol'kenshteyna [Physical Foundations of Biological Morphogenesys]* ed M V Volkenshtein (Moscow: Nauka)

Bisio G, Cartesegna M, Rubatto G and Bistolfi F 2003 Thermodynamic analysis of mechanochemical processes *Chem. Eng. Commun.* **190** 177–96

Blumenfeld L A 1981 *Problems of Biological Physics* (Springer Series in Synergetics) (Berlin: Springer)

Brillouin L 1956 *Science and Information Theory* (New York: Academic)

Cannon W B 1932 *The Wisdom of the Body* (London: W.W. Norton & Company)

Gladyshev G P 2017 Hierarchical thermodynamics: foundation of extended Darwinism *Imp. J. Interdiscip. Res.* **3** 1576–88

Grudnickij V A and Nikol'skaja I S 1977 Heat production at early stages of axolotl growth from direct and indirect calorimetry data *Ontogenesis* **8** 80–2

Gujrati P D 2018 Hierarchy of relaxation times and residual entropy: a nonequilibrium approach *Entropy* **20** 149

Kleymenov S J 1996 The mass specific energy metabolism in the early ontogenesis of animals from the data of indirect and direct calorimetry *PhD Thesis* IBR RAN, Moscow

Michaelian K 2011 Thermodynamic dissipation theory for the origin of life *Earth Syst. Dyn.* **2** 37–51

Pokrovskii V N 2018 *Econodynamics: The Theory of Social Production* 3rd edn (Dordrecht: Springer)

Popovic M 2018 *Wanted dead or alive extraterrestrial life forms (thermodynamic criterion for life is a growing open system that performs self-assembly processes)* (arXiv:1810.10389 [q-bio.OT])
Prigogine I 1961 *Introduction to Thermodynamics of Irreversible Processes* 2nd edn (New York: Interscience)
Romanovskii Y M, Stepanova N V and Chernavskiy D S 2003 *Matematicheskoe modelirovanie v biofizike (Mathematical Modelling in Biophysics)* (Moskva-Izhevsk: Institut komp'yutornykh issledovaniy)
Schrödinger E 1944 *What is Life? The Physical Aspect of the Living Cell* (Cambridge: Cambridge University Press)
Trincher K S 1965 *Biologiya i informatsiya: Elementy biologicheskoy termodinamiki [Biology And Information: Elements of Biological Thermodynamics]* 2nd edn revised and extended (Moscow: Nauka)
Turing A M 1952 The chemical basis of morphogenesis *Philos. Trans. R. Soc. Lond.* B **237** 37–72
Volkenshtein M V 1984 The essence of biological evolution *Sov. Phys. Usp.* **27** 515–37
von Stockar U and Liu J-S 1999 Does microbial life always feed on negative entropy? Thermodynamic analysis of microbial growth *Biochim. Biophys. Acta* **1412** 191–211
von Stockar U, Maskow T, Liu J-S, Marison I W and Patino R 2006 Thermodynamics of microbial growth and metabolism: an analysis of the current situation *J. Biotechnol.* **121** 517–33
Zotin A I 1972 *Thermodynamic Aspects of Developmental Biology, Monographs in Developmental Biology* vol 5 (Basel: S. Karher)
Zotin A A and Zotin A I 1997 Phenomenological theory of ontogenesis *Int. J. Dev. Biol.* **41** 917–21
Zotin A I and Zotina R S 1993 *Femenologicheskaja teorija razvitija, rosta i starenija organizmov [Phenomenological Theory of Development, Growth and Aging of Organisms]* (Moscow: Nauka)
Zotin A A and Pokrovskii V N 2018 The growth and development of living organisms from the thermodynamic point of view *Physica A: Stat. Mech. Appl.* **512** 359–66

Part III

Approaching the superorganisms

IOP Publishing

Thermodynamics of Complex Systems
Principles and applications
Vladimir N Pokrovskii

Chapter 10

Dynamics of populations

A group of similar organisms that are connected with each other by processes of co-operation and reproduction gives rise to a complex dynamical structure—*population*. A huge number of populations of animals, plants and micro-organisms can exist simultaneously on a certain territory, making up a more complex dynamical structure—*ecosystem*. Strictly speaking, one has to consider the system of biological populations together (*biocoenosis*) or even the system of biological populations together with inorganic compounds in their environment (*biogeocoenosis*), so that a mathematical model of an ecosystem could contain very many variables. However, it is convenient to start by studying simplified mathematical models of separate populations taking the other species and the environment into account through some average parameters. In a similar way, it is possible to separate and consider a few cooperating populations.

10.1 Dynamics of a separate population

Though no population can exist without the environment, from which it gets resources for living, one can, instead of the listing all details of the environment, introduce some mean characteristics and consider the dynamics of the only or a few populations, as dynamics of number of individuals in the populations. One has to think about the population as about something whole: species of population exchange messages among themselves and each species in a population plays its role in accordance with the behaviour of other species.

10.1.1 Balance equation

A change in the size of a population of organisms, which is described by variable N, is determined by four basic demographic processes: birth, death, immigration, and emigration. We omit the processes of migration (immigration and emigration) for a while, so that the change in the number of individuals in a population over a specific time interval is determined by the balance equation

doi:10.1088/978-0-7503-3451-8ch10

$$\frac{dN}{dt} = B - D. \tag{10.1}$$

In this balance equation, B and D are, respectively, the numbers of individuals born and dying per unit of time. To make equation (10.1) more useful, it is convenient to introduce the relative quantities—the fertility and mortality coefficients, correspondingly,

$$b = \frac{B}{N}, \quad d = \frac{D}{N}. \tag{10.2}$$

Now, the balance equation (10.1) can be written in another form. To describe the change of number of individuals of a the population, one can start (Lotka 1925, Volterra 1931, Murray 1989) with a simple equation of growth

$$\frac{dN}{dt} = (b - d)N. \tag{10.3}$$

The form of equation (10.3) is quite general and it can be applied to any biological population. However, coefficients b and d are determined by circumstances of the life and the environment, in which individuals of the population exist. These coefficients can depend on the number of species of the same population N and/or some extra variables, and we ought to decide what characteristics of the environment we have to attract to write the rate of population growth $b - d$.

10.1.2 Exponential growth

In the simplest case, when there are abundance of food, space and other resources, one can assume that the growth rate is a natural characteristic of the population and is constant

$$b - d = r.$$

Then, equation (10.3) determines a solution

$$N(t) = N(0) \exp rt. \tag{10.4}$$

It is possible to notice, that some populations during the limited period of time are developing according to the exponential law, which, certainly, is not be valid for the total period of development of the population. To describe empirical dependences, one can assume that the rate of growth depends on time to describe some dependence, but it is not a solution of the problem: time dependence ought to be included through some other variables.

10.1.3 The logistic growth

One can think that the amount of all necessary resources (area of habitat, food and so on) are usually restricted, so that some limiting value of number of individuals $\bar{N}$ can be supported. In this case the fertility and mortality coefficients cannot be constant and in the simplest case the growth rate can be written as a function of the number of inhabitants

$$b - d = r\left(1 - \frac{N}{\bar{N}}\right), \tag{10.5}$$

where r is intrinsic, or biological, or Malthus rate of growth.

In this case, the growth equation of the population is the known Verhulst–Pearl (Verhulst 1838, Pearl 1927; see also Murray 1989) equation

$$\frac{dN}{dt} = r\left(1 - \frac{N}{\bar{N}}\right)N. \tag{10.6}$$

In contrast to equation (10.3) with $b - d = r =$ const., equation (10.6) has a steady-state point $\bar{N}$, which is stable, because $\frac{dN}{dt} < 0$ at $N > \bar{N}$ and $\frac{dN}{dt} > 0$ at $N < \bar{N}$. Equation (10.6) describes a set of bounded trajectories, and all the trajectories approach the asymptotic line $N = \bar{N}$, as is shown in figure 10.1.

An explicit solution of equation (10.6) can be easily found, if one writes down this equation in the form

$$\frac{dN}{N} - \frac{dN}{N - \bar{N}} = r\,dt.$$

Integration of this relation, at $N < \bar{N}$ and $N > \bar{N}$, gives, correspondingly, relations

$$\ln N - \ln(\bar{N} - N) = rt + \ln A$$
$$\ln N - \ln(N - \bar{N}) = rt + \ln A$$

One can expose the quantity A through initial value $N(0)$, so that one has

$$N(t) = \frac{\bar{N}\,N(0)\exp rt}{\bar{N} - N(0) + N(0)\exp rt}.$$

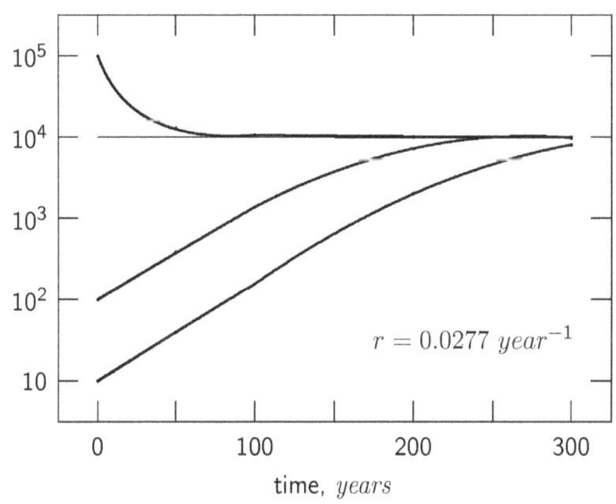

Figure 10.1. Evolution of population according to logistic growth. The trajectories, calculated according to equation (10.6), approach the line $N = \bar{N}$ from the top at $N > \bar{N}$ or from the bottom at $N < \bar{N}$. Calculated at $r = 0.027\,7$ year^{-1}, $\bar{N} = 10^4$, and different initial values $N(0)$.

This formula is valid for both $N < \bar{N}$, and $N > \bar{N}$. The equation (10.6) describes a set of the limited trajectories named by logistical curves; all trajectories at greater times when $t \to \infty$, come close to the asymptotic line $N = \bar{N}$ as is shown in figure 10.1. At small times population number changes as

$$N(t) = N(0)\left[1 + \left(1 - \frac{N(0)}{\bar{N}}\right) rt\right]. \tag{10.7}$$

At big times $t \to \infty$, in any case, the population number has limiting value $\bar{N}$.

10.1.4 Migration

Species of population are distributed over a certain area, one says, *area of habitat S*, and species of populations of equal numbers, but situated on different areas, are in different situations, because, for example, the available resources depend on the area. To formulate the law of evolution, independent of area, one ought to consider, instead of total number of species N, density of population $n = N/S$. One has to consider the density of population as a function of time and space coordinate

$$n = n(t, x, y).$$

The laws of evolution, discussed in the previous section, can be understood as local laws, while the coefficients b and d having meaning of numbers of birth and death for one specimen depend now on the point of area. The limiting number of population $\bar{N}$ is also understood as a local characteristic. Now, one has to take into account that the number density is changing not only due to births and deaths, but also due to migration of the species, that is, motion of species from one region to another in search of food and shelter.

A separate specimen of population moves over the area, and such a motion can be regarded as that similar to motion of a Brownian particle. One can introduce density of probability $W(x, y)$, so that the quantity

$$W(x, y)\,\mathrm{d}x\mathrm{d}y \tag{10.8}$$

is a probability for a specimen to move for unit of time from the origin of coordinate into the square with co-ordinates x, y. In the isotropic case, the density of probability depends on only one argument r and can be written as

$$2\pi W(r)\, r\mathrm{d}r.$$

This is the probability for a species to move for a unit of time from the origin of coordinate for the distance between r and $r + \mathrm{d}r$.

The squared mean displacement of species for unit of time

$$a^2 = 2\pi \int_0^\infty W(r)\, r^3 \mathrm{d}r \tag{10.9}$$

is named by Timofeeff-Ressovski (see Svirezhev 1987) as *the radius of individual activity*. If this quantity is much less than typical scale of the area ($a \ll R$) (assuming the unit of time is close to the time of living), one can consider the population to be homogeneous inside the area. Otherwise, one has to take into account the size of area explicitly.

The density of population in time $t + \Delta$ in point x, y can be expressed through the density of population in time t in all points of the area of habitat, which we consider in the following form

$$\begin{aligned} n(t + \Delta t, x, y) = {} & n(t, x, y) \\ & + \Delta t \int_{-\infty}^{+\infty} [n(t, x', y') - n(t, x, y)] \\ & \times W(x' - x, y' - y)\, \mathrm{d}x'\mathrm{d}y'. \end{aligned}$$

The integrated area is considered as infinite, assuming that the radius of individual activity is much less than typical scale of the area.

Assuming that contributions from the points x', y' are the more significant the less the distance between this point and the point x, y, one can account for only the first terms in expansion

$$\begin{aligned} n(t, x', y') - n(t, x, y) = {} & \frac{\partial n}{\partial x}(x' - x) + \frac{\partial n}{\partial y}(y' - y) \\ & + \frac{\partial^2 n}{\partial x^2}(x' - x)^2 + 2\frac{\partial^2 n}{\partial x \partial y}(x' - x)(y' - y) \\ & + \frac{\partial^2 n}{\partial y^2}(y' - y)^2 + \cdots. \end{aligned}$$

Making use of property of symmetry of function of probability, one can get the relation

$$\Delta n = \frac{1}{4}a^2\left(\frac{\partial^2 n}{\partial x^2} + \frac{\partial^2 n}{\partial y^2}\right)\Delta t. \tag{10.10}$$

Now, one can get the migration equation

$$\frac{\partial n}{\partial t} = D\left(\frac{\partial^2 n}{\partial x^2} + \frac{\partial^2 n}{\partial y^2}\right) \tag{10.11}$$

where $D = a^2/4$ is a coefficient of migration. The above equation is the equation of diffusion of a Brownian particle on the flat surface and coefficient of migration is coefficient of diffusion.

To the right side of equation (10.11), one can add a source term, which is connected with births and deaths of species, for example, in the form written in equation (10.6)

$$\frac{\partial n}{\partial t} - D\left(\frac{\partial^2 n}{\partial x^2} + \frac{\partial^2 n}{\partial y^2}\right) = r\left(1 - \frac{N}{\bar{N}}\right)N. \tag{10.12}$$

It is possible to assume also, that there is some delay of migration in the form, used, for example, in the work of the Fort and Méndez (1999).

To describe the propagation of species from a point source, one can write the equation of migration in the form

$$\frac{\partial n}{\partial t} - D\frac{\partial}{\partial r}\left(r\frac{\partial n}{\partial r}\right) = n(0)\,\delta(r) \tag{10.13}$$

where r is the radial coordinate. The solution of the above equation

$$n(t,\, r) = \frac{n(0)}{(4\pi Dt)^{3/2}}\exp\left(-\frac{r^2}{4Dt}\right)$$

describes the propagation of species from an initial point source.

10.1.5 A wave in logistic population

Let us consider, following the work of Svirezhev (1987), propagation of population, for simplicity, over a one-dimensional area, being interested in displacement of the boarder of the population. Equation (10.12) for this case can be written in the form

$$\frac{\partial n}{\partial t} - D\frac{\partial^2 n}{\partial x^2} = F(n) \tag{10.14}$$

where $F(n)$ is a function of local growth of density of population with a certain limit value $\bar{n}$, while

$$F(0) = F(\bar{n}) = 0, \quad 0 < \bar{n} < \infty$$
$$F'(n) < F'(0) = r > 0, \quad n > 0.$$

The function can, for example, have a form shown in equation (10.6).

To investigate motion of the border, let us present the density of population, which depends on both time t and coordinate x, as a function of the only argument

$$n(t,\, x) = n(x + vt) = n(\xi). \tag{10.15}$$

This presentation assumes that the function of population density does not change its form, and the population propagates to the left side with velocity v. The new border conditions are valid now

$$n(-\infty) = 0, \quad n(+\infty) = \bar{n}.$$

One can use function (10.15), to rewrite equation (10.14) in the form

$$v\,\frac{\partial n}{\partial \xi} = D\frac{\partial^2 n}{\partial \xi^2} + F(n). \tag{10.16}$$

The equation can be written in the form of the system of the two first-order equations

$$\frac{\partial n}{\partial \xi} = p,$$
$$\frac{\partial p}{\partial \xi} = \frac{v}{D} p - \frac{1}{D} F(n).$$

The system has the two steady-state points

$$p = 0,\ n = 0;$$
$$p = 0,\ n = \bar{n}.$$

The equation of trajectory on the phase plane (n, p) has the following form

$$\frac{\partial p}{\partial n} = \frac{v}{D} - \frac{1}{p\,D} F(n). \tag{10.17}$$

These results are apparently the consequences of the simplest version of description, which can be generalized in a number of ways.

10.1.6 The composition of a population

The specimens of every population are different in age, gender and many other features. The behaviour of species of different ages and gender are also different, so that, to describe the evolution of population in more detail (Polujektov 1974), it is necessary to distinguish numbers of individuals of certain groups, for example, to introduce the numbers of individuals for male and female, $N^{♂}$ and $N^{♀}$, accordingly, so that the total number of population is

$$N(t) = N^{♂}(t) + N^{♀}(t).$$

Further, for a more accurate description it is necessary to introduce age of the individual τ and density of numbers of individuals for males and females, $N^{♂}(t, \tau)$ and $N^{♀}(t, \tau)$, accordingly, so, for example, $N^{♂}(t, \tau)\,\mathrm{d}\tau$ is a number of males aged from τ up to $\tau + \mathrm{d}\tau$ at point in time t. Certainly, the parities are valid

$$N^{♂}(t) = \int_0^{\infty} N^{♂}(t, \tau)\,\mathrm{d}\tau, \quad N^{♀}(t) = \int_0^{\infty} N^{♀}(t, \tau)\,\mathrm{d}\tau.$$

The variation of numbers due to birth and death of individuals is described, apparently, by the equations

$$\begin{aligned} \frac{\mathrm{d}N^{♂}(t, \tau)}{\mathrm{d}t} &= -\,d^{♂}(t, \tau) N^{♂}(t, \tau), \\ \frac{\mathrm{d}N^{♀}(t, \tau)}{\mathrm{d}t} &= -\,d^{♀}(t, \tau) N^{♀}(t, \tau) \end{aligned} \tag{10.18}$$

with the initial conditions

$$N^{\male}(t, 0) = \int_0^{\infty} b^{\male}(t, \tau')N^{\female}(t, \tau')\mathrm{d}\tau',$$
$$N^{\female}(t, 0) = \int_0^{\infty} b^{\female}(t, \tau')N^{\female}(t, \tau')\mathrm{d}\tau'.$$

In these relations, factors of birth rate and death rate, depending on gender and age, are introduced.

To return to a more rough description, one can integrate equations (10.18) on a variable τ from 0 up to ∞, taking into account the conditions recorded above, which defines

$$\begin{aligned}\frac{\mathrm{d}N^{\male}(t)}{\mathrm{d}t} &= \int_0^{\infty} b^{\male}(t, \tau)N^{\female}(t, \tau)\mathrm{d}\tau - \int_0^{\infty} d^{\male}(t, \tau)N^{\male}(t, \tau)\mathrm{d}\tau,\\ \frac{\mathrm{d}N^{\female}(t)}{\mathrm{d}t} &= \int_0^{\infty} b^{\female}(t, \tau)N^{\female}(t, \tau)\mathrm{d}\tau - \int_0^{\infty} d^{\female}(t, \tau)N^{\female}(t, \tau)\mathrm{d}\tau.\end{aligned} \tag{10.19}$$

In the more general case, each specimen of population can be ascribed a few characteristics, so that each individual requires a label with a few indexes, which allows one to consider processes of transition of the individual from one group into another and to investigate features of emergence of groups.

10.2 Interacting populations

In some cases, a population, which is an object of our interest, is very closely connected with other populations, and one has to investigate the evolution of the co-existing populations to understand the behaviour of a chosen one. In the simplest case, one can consider two populations, as a simplest model of biocoenosis. The characteristic of each population is the number of individuals, and dynamics of the population is considered as dynamics of number of individuals of the population. In this case, one can write a general form of equation

$$\frac{\mathrm{d}N_1}{\mathrm{d}t} = F_1(N_1, N_2)$$
$$\frac{\mathrm{d}N_2}{\mathrm{d}t} = F_2(N_1, N_2).$$

The behaviour of populations can be quite different at different specifications of the right-side functions of the above equations.

10.2.1 The model of competing populations

If the specimens of two populations use one and the same food, the populations compete, and the growth rate of each population depends on the population numbers of each population, that is,

$$\frac{dN_1}{dt} = (r_1 - c_{11}N_1 - c_{12}N_2)N_1$$
$$\frac{dN_2}{dt} = (r_2 - c_{21}N_1 - c_{22}N_2)N_2 \qquad (10.20)$$

where r_1 and r_2 are the intrinsic growth rate for either population, $c_{jk} > 0$ are coefficients of interaction and self-interaction. In the case, when the amount of food is unrestricted, the numbers of populations change according to exponential growth with their own coefficients of growth r_1 and r_2, but in the cases, when the mutual influence is essential, the behaviour of population becomes more complicated.

For steady-state situations, one has the systems of equations for the population numbers

$$(r_1 - c_{11}N_1 - c_{12}N_2)N_1 = 0,$$
$$(r_2 - c_{21}N_1 - c_{22}N_2)N_2 = 0. \qquad (10.21)$$

The system determines, apart from trivial steady-state point $N_1 = 0$, $N_2 = 0$, three stationary points

$$N_1 = 0, \qquad N_2 = \frac{r_2}{c_{22}},$$
$$N_1 = \frac{r_1}{c_{11}}, \qquad N_2 = 0, \qquad (10.22)$$
$$N_1 = \frac{r_1c_{22} - r_2c_{12}}{c_{11}c_{22} - c_{12}c_{21}}, \qquad N_2 = \frac{r_1c_{11} - r_2c_{21}}{c_{11}c_{22} - c_{12}c_{21}}$$

It is assumed, that the coefficients c_{jk} have such values that the denominator of the above expressions is not equal to zero.

Figure 10.2 shows trajectories of evolution, defined by equations (10.21) at initial population numbers $N_1(0)$ and $N_2(0)$. The steady-state points can be also seen. To

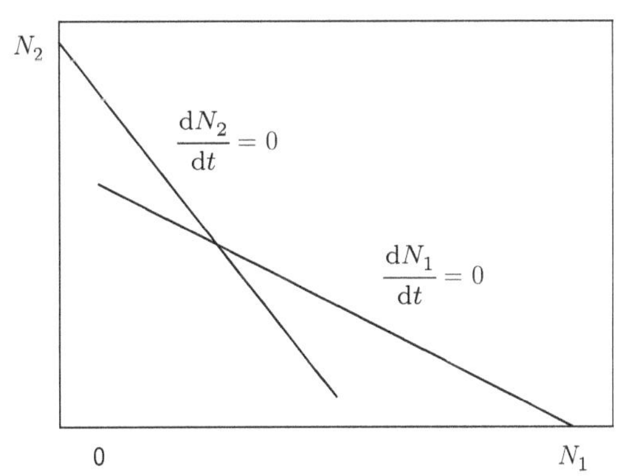

Figure 10.2. Phase space of competing populations. Points of intersection of straight lines with each other and with coordinate axes are stationary points, the stability of which depends on values of factors c_{jk}.

investigate the behaviour of the system in the vicinity of the steady-state points (10.22), one can present the population numbers as

$$N_1 = N_1^{(s)}(1 + x_2), \quad N_2 = N_2^{(s)}(1 + x_2), \tag{10.23}$$

where $N_1^{(s)}$ and $N_1^{(s)}$ are values of population numbers in one of the stationary points ($s = 1, 2, 3$). The small, as compared with unity, quantities x_1 and x_2 are obeying to the equation

$$\frac{dx_1}{dt} = - c_{11}N_1^{(s)}x_1 - c_{12}N_2^{(s)}x_2$$
$$\frac{dx_2}{dt} = - c_{21}N_1^{(s)}x_1 - c_{22}N_2^{(s)}x_2.$$

Considering three cases separately, we find, that stability of the system in stationary points depends on signs on factors c_{jk}. It is possible to make sure, that, depending on initial values and values of the parameters, one of the populations survives, or two populations can coexist in the case, if the last stationary point is stable. The condition of stability appears a condition of coexistence. Otherwise, co-existence of two populations appears impossible.

10.2.2 The predator–prey model

Italian mathematician Vito Volterra attained the main suggestions about the coexistence of two populations, when he considered a huge volume of data, witnessing pulsations of some species of fish in the Adriatic sea. These pulsations had a time behaviour close to periodic, and there were no reasons to connect these pulsation with any geographic or climatic changes. These observations have laid down in a basis of the model predator–prey (Volterra 1931).

Consider two interacting populations, the species of one of which (prey) is food for the other (predator). Let N_1 and N_2 be the population numbers of prey and predator. One can assume, for the beginning, that resources of the first population (prey) are not restricted, but the species of the second population (predator) apparently cannot live without food provided by the species of the other population, so that, if there is no direct interaction between species of the two populations, one can write

$$\frac{dN_1}{dt} = r_1N_1, \quad r_1 > 0,$$
$$\frac{dN_2}{dt} = - r_2N_2, \quad r_2 > 0.$$

where r_1 and r_2 are the growth rate for either population in the absence of interaction. In this case, the population number of prey grows, and the population number of predator decays, according to exponential dependence with their own coefficients of growth r_1 and decay r_2. To take into account the interaction of the populations, non-linear terms ought to be added to equations

$$\frac{dN_1}{dt} = r_1N_1 - \gamma_1N_1N_2, \quad \gamma_1 > 0,$$
$$\frac{dN_2}{dt} = -r_2N_2 + \gamma_2N_1N_2, \quad \gamma_2 > 0. \tag{10.24}$$

In the cases when the mutual influence is essential, the growth rate decreases for the first population and increases for the second population.

For steady-state situations, one has the systems of equations

$$r_1N_1 - \gamma_1N_1N_2 = 0,$$
$$-r_2N_2 + \gamma_2N_1N_2 = 0,$$

so that, apart from trivial steady-state point $N_1 = 0$, $N_2 = 0$, one has only one stationary point

$$N_1 = \frac{r_2}{\gamma_2}, \quad N_2 = \frac{r_1}{\gamma_1}. \tag{10.25}$$

An equation for the trajectory of evolution can be easily found from the system of equations (10.24)

$$\frac{dN_1}{dN_2} = \frac{N_1}{N_2}\frac{r_1 - \gamma_1N_2}{-r_2 + \gamma_2N_1}. \tag{10.26}$$

One can separate the variables in the above equation

$$-r_2\frac{dN_1}{N_1} + \gamma_2 \, dN_1 = r_1\frac{dN_2}{N_2} - \gamma_1 \, dN_2$$

and find the equation of the trajectory of motion of the system on the phase surface

$$\gamma_2 N_1 - r_2 \ln N_1 = r_1 \ln N_2 - \gamma_1 N_2 + \ln A, \tag{10.27}$$

where A is the integration constant, which is determined by initial values.

To consider, firstly, the trajectory of evolution close to the steady-state point (10.25), one can introduce small, as compared with unity, variables x_1 and x_2 and present the population numbers as

$$N_1 = \frac{r_2}{\gamma_2}(1 + x_1), \quad N_2 = \frac{r_1}{\gamma_1}(1 + x_2). \tag{10.28}$$

Now, the above equation of trajectory can be simplified. One can use an approximate equation

$$\ln(1 + x) \approx x - \frac{1}{2}x^2,$$

to find

$$r_2x_1^2 + r_1x_2^2 = 2(\ln A - \ln A_0),$$

$$\ln A_0 = r_1\left(1 - \ln\frac{r_2}{\gamma_2}\right) + r_2\left(1 - \ln\frac{r_1}{\gamma_1}\right),$$

where A is the value of the integration constant for an arbitrary trajectory, whereas A_0 is the value of the integration constant for the trajectory, which comes through the steady-state point, so that the right-hand side of the above equation depends only on initial values $x_1(0)$ and $x_2(0)$.

One can see that, in the vicinity of steady-state point, the system is moving along a closed curve—the ellipse. The trajectories of evolution are depicted for initial values $N_1(0)$ and $N_2(0)$ on the plot in figure 10.3. One can show by different ways that at arbitrary values of initial values, the trajectory is a close curve as well. To determine the direction of motion, one can simplify equations (10.24) to obtain the equations

$$\frac{dx_1}{dt} = -r_1x_2, \qquad \frac{dx_2}{dt} = r_2x_1. \tag{10.29}$$

10.2.3 Dynamics and thermodynamics of biogeocoenosis

After having discussed some models in the previous sections, it is becoming clear how a mathematical model of evolution of any number of interacting populations can be written. Let n_j be the density of population number with label j. Then, a set of interdependent equations for m interacting populations can be written, taking also processes of migration into account, as

$$\frac{\partial n_j}{\partial t} = \left(r_j - \sum_{k=1}^{m} c_{jk}n_k\right)n_j + D_j\left(\frac{\partial^2 n_j}{\partial x^2} + \frac{\partial^2 n_j}{\partial y^2}\right), \quad j = 1, 2, \ldots, m \tag{10.30}$$

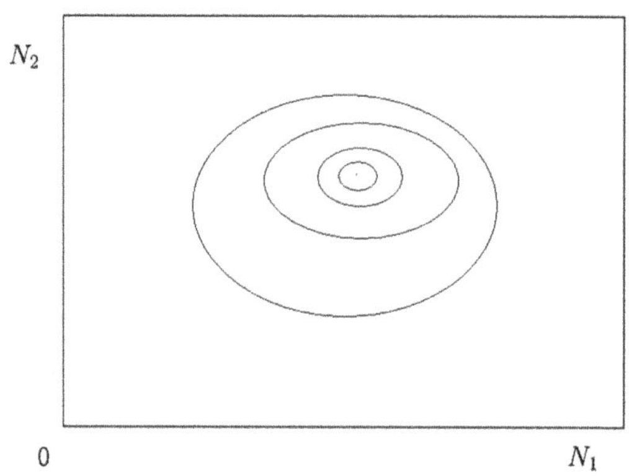

Figure 10.3. Evolution of the system predator–prey. Trajectories of the system represent a closed curve surrounding a stationary point. At small deviations from a stationary point curves appear as ellipses.

where D_j is a coefficient of migration of species of population j. The coefficients r_j and c_{jk} have arbitrary signs, so that the equations considered in previous sections, (10.20) and (10.24), are particular cases of system (10.30). The system is capable of describing various features of behaviour of cooperating populations that were observed in a simple case $m = 2$, studied in the previous sections.

The system of the equations (10.30) for cooperating populations describes a complex picture of development of a *biocoenosis* in a rather coarse approximation: each population is characterized only by density of number of individuals. For a more detailed description of individuals of each population indexes of gender, age and other characteristics of the species can be attributed, and appropriating variables are introduced.

We shall get a more complete description, if we include in the description variables showing availability of organic and inorganic substances from an environment: air, water, minerals and other. At inclusion of mineral components, the considered system is referred to as *biogeocoenosis*. Apparently, there is a set of the special cases, which are of research interest: the investigation of various particular cases is the contents of work of scholars, the papers of whom are published in various scientific journals.

From the point of view of thermodynamics, it is possible to consider both *biocoenosis*, and *biogeocoenosis* as an open thermodynamic system, which, however, is not allocated by constant borders; interoperability with the environment occurs on the level of individuals. Nevertheless, the ecosystem (biocoenosis or biogeocoenosis) as well as a separate organism, described in the ninth chapter, can be characterized with thermodynamic functions of state that can be presented as the sum of thermodynamic characteristics of individuals with possible additives from the effect of interoperability of individuals

$$G = NG_{\text{ind}} + G_{\text{soc}}, \qquad S = NS_{\text{ind}} + S_{\text{soc}}. \tag{10.31}$$

Here, the symbols G_{ind} and S_{ind} are Gibbs free energy and entropy of a separate individual; we believe that variation of these quantities is connected not only with course of chemical reactions and delivery of substances, as shown in definitions (9.1) and (9.2), but also with other possible processes. Symbols G_{soc} and S_{soc} designate thermodynamic quantities that are defined by joint, social work of the individuals. These quantities, apparently, are negligible for all populations, except for a human population. Thermodynamic characteristics G and S depend on temperature and pressure that appear insignificant; the current non-equilibrium state of the system is essentially defined by internal variables. Among the set of internal variables we find density of distribution of individuals of various populations on the area n_j ($j = 1, 2, \ldots$); these variables are describing (from the point of view of thermodynamics) large-scale structures.

The existence of every biocoenosis or biogeocoenosis is based on energy coming into the system from the environment through the food and organisms of individuals, finally, from the Sun through various transformations: photosynthesis of plants, various trophic circuits and so on. After numerous chemical transformations and spatial redistributions some substances are removed from the organism. Capacity to take energy from the environment is a feature of an ecosystem. The natural mechanism for extraction of energy from the environment through food and organisms of individuals defines a *biologically organized stream of energy. The power principle of evolution*, which has been stated by many researchers (Lotka 1925, Petchurkin 1982, Odum 1996), states that the direction of evolution of a population (ecosystem) is such that it provides greater capacity to take energy from the environment. Those populations and their associations (ecosystems), capable of taking a lot of energy from the environment, have the advantage for survival.

10.3 Peculiarities of development of human population

Human society should be considered, first of all, as one of biological populations, obeying the laws that are mutual for all populations. However, there is something that distinguishes a human population from others and allows us to separate out this population for special consideration. People are not satisfied with that food and those refuges that they find in the natural environment, they have created and are supporting systems of production of food, clothes, refuges and other necessary things, so that the human population appears to be a unique population. Since Palaeolithic times clothing, shelter and fuel have become necessities of life almost as fundamental as food itself. The development of the human population is based on controlled use of energy. Certainly, it is necessary to assume, that origin of these features of the human population is in biological features of humans, in the features of the structure of their brains, capacity of a human to reflect. The last means that the human thinks not only of things, which he perceives with the sense organs, but also about the thoughts on those sensations.

10.3.1 Empirics of population growth

Discovery and study of the remains of ancient humans allows us to state that humans appeared in Africa about three millions years ago (Leakey 1981), and since that remote time extended into almost all territories of the Earth. Archeological evidence coupled with analysis allows one to restore the alleged history of the tumbling movement of hominids from Africa though straits and ridges from 2 million until 500 thousand years ago by a computer model (Mithen and Reed 2002). The available current methods of studying of mutations in a Y-chromosome of remains of ancient people allow a clear picture of evolution and movement of humans (Klyosov and Tyunyaev 2010). These methods have allowed us to show, that all people living now are descendants of a small homogeneous group of people which survived after a planetary accident that happened no later than 64 thousand years ago; this figure—the age of an ancestor was estimated by methods of DNA-genealogy (Klyosov and Tyunyaev 2010). During the subsequent millennia, humans of a modern type, superseding probably, previous types of humans, occupied Europe and Asia (Klyosov and Tyunyaev 2010). The population of humans has acquired such great abilities to live in any climate zone of the Earth that the numbers are greatly enlarged.

Assessments of the number of the population N on the Earth according to various sources are shown in figure 10.4. Fast growth occurred in some instants of history

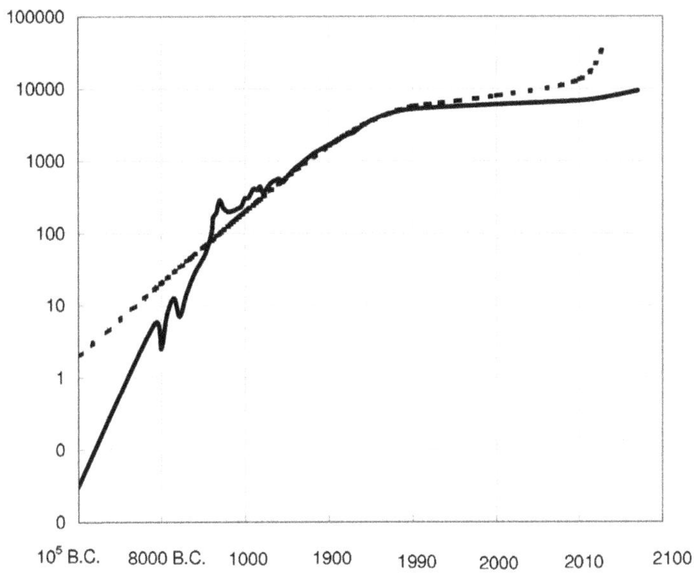

Figure 10.4. Growth of the world population. The population of the Earth in millions. The solid line represents average values of available estimates of the population number (http://www.census.gov/ipc/www/worldhis.html). The dashed line is drawn according to equation (10.32).

designated as a cultural revolution (nearby 10^6 years back), agricultural revolution (nearby 10^4 years back) and industrial revolution, which began in the 17th–18th Centuries. During the last four thousand years the population of the Earth has increased from 30 million up to almost 7 billion. Growth of number of humans during four millennia can be approximated (Shklovsky 1987, section 25) with the hyperbolic law

$$N(T) = \frac{2 \times 10^{11}}{2030 - T}, \tag{10.32}$$

where T is time in years of our chronology. Dependence (10.32) is reproduced in figure 10.4. Certainly, as Shklovsky remarks, it is not necessary to give to the point at $T = 2030$ some sense or profound meaning: after 2000, from comparison of a curve with empirical data, the approximation apparently does not work.

The huge growth of the human population number testifies to the evolutionary success of the population, which is connected with special features of the human population in comparison with other biological populations occupying the Earth. Apparently, it is impossible to explain, without referring to a social production system—a way of adaptation of humans to conditions of existence, an increase of the human population from a very small group one million years ago up to approximately 6 billion in 2000.

10.3.2 The laws of growth of a human population

To describe the changes of numbers of the human population, it is convenient to start with the simple balance equation (10.3), which, as was noted, is applicable to any biological population. Peculiarities of evolution of the human population should be connected with circumstances of life of individuals of the population, which are influencing the factors of birth rate and death rate, accordingly, b and d.

While investigating development of the human population, Malthus (1798) found[1], that, if the growth of the population is not restrained, number of the population doubles each 25 years, which allowed Malthus to state that in these conditions exponential growth (10.4) is realized with value of growth rate

$$r \approx 0.027\,7\ \text{year}^{-1}, \tag{10.33}$$

This quantity defines time $1/r \approx 36.1$ that is the time of increasing the population number in $e \approx 2.72$ times. The Malthus constant r defines some time scale that is close to the life time of a generation.

Since Malthus' times, it becomes clear, that availability of resources restricted the human population growth, and investigation of the nature and character of restrictions has appeared central and fruitful in research of the problem of growth of the human population. The convenient tool for this case has appeared to be the mathematical model (described in section 10.1.3) established for biological populations (Verhulst 1838, Pearl 1927) that has found widespread use in the description of features of growth of the human population (Turchin 2003, 2009, Nefedov 2007). Equation (10.6) can serve as a convenient initial point for interpretation of dynamics of evolution of human population (Pokrovskii 2018, chapter 12). According to this equation, population number does not exceed the limiting value $\bar{N}$ that changes for the human population over time. Possible value of the population number $\bar{N}(t)$ is determined by the level of technological and organizational progress which allows a person convenience in getting food, arrangement of dwelling, protection against illnesses and other advantages against other populations.

The person lives in the environment that is designed on the demands. Water, heat, and food come into dwellings. All this, including non-material values (systems of

[1] In the first chapter of the first edition of his book, Malthus writes: 'It has been universally remarked that all new colonies settled in healthy countries, where there was plenty of room and food, have constantly increased with astonishing rapidity in their population. Some of the colonies from ancient Greece, in no very long period, more than equaled their parent states in numbers and strength. And not to dwell on remote instances, the European settlements in the new world bear ample testimony to the truth of a remark, which, indeed, has never, that I know of, been doubted. A plenty of rich land, to be had for little or nothing, is so powerful a cause of population as to overcome all other obstacles. … The consequence of these favourable circumstances united was a rapidity of increase probably without parallel in history. Throughout all the northern colonies, the population was found to double itself in twenty-five years. … In New Jersey the period of doubling appeared to be twenty-two years; and in Rhode island still less. In the back settlements, where the inhabitants applied themselves solely to agriculture, and luxury was not known, they were found to double their own number in fifteen years, a most extraordinary instance of increase. Along the sea coast, which would naturally be first inhabited, the period of doubling was about thirty-five years; and in some of the maritime towns, the population was absolutely at a stand.

knowledge, organizational schemes, ethical rules, works of art and so on), represents social wealth, which is considered in conditional monetary units and is designated by the symbol W. The set of all things, comprising social wealth, can be defined as *an artificial environment* that is the important attribute of human existence and should be taken into consideration. The observed growth of human population on the Earth can be explained quantitatively, if quality of the close environment of human beings measured by the social wealth is taken into account.

Growth of the human population should be considered as the result of multiplication of individuals and simultaneous growth of the public abundance increasing limiting value of population number $\bar{N}$. A quantitative measure of the gained standard of living can serve a total size of the abundance falling for one person W/N, and it is possible to assumed that value of the quantity $\bar{N}$ increases with increase of social wealth per capita W/N, so that there is a function

$$\bar{N} = Nf\left(\frac{W}{N}\right). \tag{10.34}$$

During ancient times growth rate of social wealth was very small, and number of population N was close to the possible number $\bar{N}$; the ratio $N/\bar{N}$ was close to unity, and the deviation of this size from unit can be connected linearly with social wealth per single person (Pokrovskii 1999). For a more detailed description, it is necessary to consider separately behaviour of coefficients of birth rate and death rate, each of which is assumed to depend on social wealth per single person W/N. According to observations, it is possible to believe (Pokrovskii 2018, chapter 12) that influence of wellbeing leads to reduction of both the birth rate and mortality rate coefficients in comparison with some natural values, which are considered equal, assuming, that in the natural conditions the population number does not change. The growth rate of population $b - d$ can show non-monotonic dependence that allows one to explain observable fluctuations of the growth rate of humans.

In detailed study of the problem, it is necessary to take into account that the total social wealth is distributed non-uniformly, both inside of a population, and in relation to various local populations. Besides, it is necessary to consider that at the present stage of progress of a social production system, a part of national social wealth W, accounted for in statistical reports, not only does not assist survival of individuals of a population, but also is sometimes directly aimed at destruction of individuals. All this leads to that the part of national social wealth should be excluded from consideration: average value W/N mismatches true value of wellbeing for the majority of individuals.

The above-stated circumstances determine the general direction of progress of the human population through the simple mechanism of evolution: things and ideas, which are necessary or useful to a survival of humans, are kept and stored. In the elementary approximation, the problem of the growth of human population reduces to a problem about mutual growth of the population and production system.

The investigation of evolution of local populations in the past is very important for understanding laws of growth. It has been noticed that growth of local populations, as a rule, has a cyclic character, and to explain this, researchers involve other variables connected with the description of structure of a society and redistribution of resources. Simple models of structural-demographic theory allow one to describe increase and fall of local human populations (Nefedov 2007, Turchin 2003, 2009). It is possible to think that the discovered laws can be used for the description of the future progress of mankind, on the basis of what appropriate social mechanism of regulation of growth of the population can be proposed.

10.3.3 Thermodynamics of the human population

The human population, as well as any biological population (see section 10.2.3), can be considered as a thermodynamic system situated in some environment and cooperating with it at the level of individuals. The selected system exchanges substance and energy with the environment, and the system should be classified as the open thermodynamic system existing in a non-equilibrium state.

Gibbs free energy and entropy of any population can be represented, by equation (10.31), as a sum of thermodynamic characteristics of individuals, G_{ind} and S_{ind}, with possible additives from the effect of interoperability of individuals, G_{soc} and S_{soc}, that is

$$G = NG_{\text{ind}} + G_{\text{soc}}, \qquad S = NS_{\text{ind}} + S_{\text{soc}}. \tag{10.35}$$

It was noticed already, that for all populations, except for the human population, in definitions of thermodynamic functions (10.35), the last terms could be neglected. For the human population, the contribution from mutual, social work of individuals, G_{soc} and S_{soc} in thermodynamic characteristics appears crucial. Existence of the human population, as well as any biological population, is based on the energy coming into the system from the environment, but the human population in addition to *biologically organized stream of energy* has *socially organized stream of energy*. Except for the natural biological mechanism of extraction of energy from the environment through food and organisms of individuals, the human has learned to take for himself energy through the production system, which plays a role of a mechanism involving energy from various sources, among them the remains of former biospheres: wood, coal, oil; a direct and indirect solar power in the form of streams of air and water; energy of splitting and synthesis of atoms and nuclei.

The stream of energy into the system should be considered as a stream of enthalpy $H = G + TS$ that, similarly to quantities (10.35), can be conveniently presented in the form of the sum of individual contributions and the additive term from effect of interoperability of individuals

$$H = NH_{\text{ind}} + H_{\text{soc}}. \tag{10.36}$$

This definition and the assumption that properties of the separate individual do not change ($H_{\text{ind}} = \text{const.}$), due to the identity

$$\frac{\mathrm{d}}{\mathrm{d}t}\left(\frac{H}{N}\right) = \frac{1}{N}\frac{\mathrm{d}H}{\mathrm{d}t} - \frac{H}{N^2}\frac{\mathrm{d}N}{\mathrm{d}t},$$

allow us to define the growth rate of the population

$$\frac{1}{N}\frac{\mathrm{d}N}{\mathrm{d}t} = \frac{1}{H_{\mathrm{soc}}}\frac{\mathrm{d}H_{\mathrm{soc}}}{\mathrm{d}t} - \frac{N}{H_{\mathrm{soc}}}\frac{\mathrm{d}}{\mathrm{d}t}\left(\frac{H_{\mathrm{soc}}}{N}\right). \tag{10.37}$$

The growth rate of the population is determined by the growth rate of the socially-involved energy (the first term in the right part of the above equation) and the growth rate of the fraction of the socially-involved energy on one individual (the second term). The last term affects the 'qualitative' variation of individuals, while the first term shows quantitative growth of the population at constant 'quality' of the individuals.

The stream of energy $\frac{\mathrm{d}H_{\mathrm{soc}}}{\mathrm{d}t}$ is connected with the social production system. Energy of external sources through various adaptations is used for transformation of substances of the natural environment into things of the artificial environment, creating complexity useful to people. Part of the energy is lost and dissipates in the form of heat energy, so that the pure incoming stream of energy can be defined as a power assessment of efforts of humans and work of the production equipment, $P + L$ (see section 8.4.2). Taking into account also interpretation of enthalpy as $H_{\mathrm{soc}} \approx T\ S_{\mathrm{soc}}$, we record the expression for the growth rate of the population through thermodynamic quantities

$$\frac{1}{N}\frac{\mathrm{d}N}{\mathrm{d}t} = \frac{1}{T\ S_{\mathrm{soc}}}(P + L) - \frac{N}{S_{\mathrm{soc}}}\frac{\mathrm{d}}{\mathrm{d}t}\frac{S_{\mathrm{soc}}}{N}. \tag{10.38}$$

The change of the population number is connected with both an increase of the stream of energy into the system (the first term) and with qualitative variations of the system (last term).

Equation (10.38) represents the formulation *power principle of evolution*. The investigation of dynamics of production shows (Pokrovskii 2003, 2018), that a principle on which production can be developed is stated as a principle of the maximal devouring of available resources. In fact, this principle of evolution is a power principle of progress: the trajectory of development of the system is defined by aspiration of the system to use the greatest quantity of available energy. More and more energy is used by the human population through improvements of technology. The previous history of mankind confirms fairness of a power principle of evolution. The possession of a huge quantity of energy allows a human population to survive in all climatic zones of the Earth and to extend throughout the globe. Besides, it is possible to see from the history of mankind, that the nations which possessed methods of use of available energy, got an advantage before other nations. It is possible to address classic examples: the industrial revolution and prosperity of Great Britain began with the time of the invention of steam machines, which allowed the use of chemical energy of coal stored in the Earth. World history can be rewritten as a history of struggle for control over streams of energy.

References

Fort J and Méndez V 1999 Time-delayed theory of the neolithic transition in Europe *Phys. Rev. Lett.* **82** 867–70

Klyosov A A and Tyunyaev A A 2010 *Proischozhdenie cheloveka: po dannym arheologii, antropologii i DNK-geologii [Origin of the Humans: According to the Data by Archaeology, Anthropology, and the DNA Geneology]* (Boston: Belye al'vy)

Leakey R E 1981 *The Making of Mankind* (Amsterdam: Elsevier)

Lotka A J 1925 *Elements of Physical Biology* (Baltimore, MD: Williams and Wilkins)

Malthus T R 1798 *An Essay on the Principles of Population, as it Affects the Future Improvement of Society* (London: J Johnson)

Mithen S and Reed M 2002 Stepping out: a computer simulation of hominid dispersal from Africa *J. Hum. Evol.* **43** 433–62

Murray J D 1989 *Mathematical Biology* (Berlin: Springer)

Nefedov S A 2007 *Koncepcija demograficheskih tsiklov [The Doctrine of the Demographic Cycles]* (Ekaterinburg: Izdatel'stvo UGGU)

Odum H T 1996 *Environmental Accounting: Emergy and Environmental Decision Making* (New York: Wiley)

Pearl R 1927 The growth of population *Q. Rev. Biol.* **2** 532–48

Pechurkin N S 1982 *Jenergeticheskie aspekty nadorganizmennyh system [Energy Aspects of supra-organismal systems]* (Novosibirsk: Nauka)

Polujektov R A (ed) 1974 *Dinamicheskaja teorija biologicheskih populjacij [The Dynamic Theory of Biological Populations]* (Moscow: Nauka-Glavnaja redakcija fiziko-matematicheskoj literatury)

Pokrovskii V N 1999 *Physical Principles in the Theory of Economic Growth* (Aldershot: Ashgate Publishing)

Pokrovskii V N 2003 Energy in the theory of production *Energy* **28** 769–88

Pokrovskii V N 2018 *Econodynamics: The Theory of Social Production* 3rd edn (Dordrecht: Springer)

Svirezhev J M 1987 *Nelinejnye volny, dissipativnye struktury i katastrofy v jekologii [Nonlinear waves, dissipative structures, and catastrophics in ecology]* (Moscow: Nauka-Glavnaja redakcija fiziko-matematicheskoj literatury)

Shklovsky I S 1987 *Vselennaja, zhizn', razum [The Universe, Life, Mind]* ed N S Kardashev and V Y Moroz 6th edn revised and enlarged (Moscow: Nauka-Glavnaja redakcija fiziko-matematicheskoj literatury)

Turchin P 2003 *Historical Dynamics: Why States Rise and Fall* (Princeton, NJ: Princeton University Press)

Turchin P 2009 Long-term population cycles in human societies In: *The Year in Ecology and Conservation Biology* — a special issue of *Ann. N. Y. Acad. Sci.* **1162** 1–17

Verhulst P F 1838 Notice sur la loi que la population suit dans son accroissement *Corr. Math. Phys.* **10** 113–21

Volterra V 1931 *Lesons sur la mathematique de la lutte pour la vie* (Paris: Marcel Brelot)

Chapter 11

Problems of sociodynamics

The structure and laws of development of social formations can be presented owing to investigations by Cont, Durkheim, Marx, Weber, Parsons and many other researchers (Turner 1974). They succeeded in formulation of some concepts (categories) that allow us to describe social structures and laws of their variation, as well as the attitude of individuals to their roles in society, that have comprised the content of sociodynamics. Society represents a set of humans, who cooperate among themselves in many respects and various ways, so that society is perceived and operates as the whole. Considering the population as a thermodynamic system, we can, as well as with any biological population, attribute thermodynamic characteristics to the human population, but from the entire set of values, only entropy is sometimes used to characterize the complexity of society. All the quantities, which are used to describe the structure and functionality of society, should be considered as thermodynamic internal variables.

11.1 Architecture of social organization

Communities of people have various names: a population, society, nation, ethnos⋯. The known Russian philosopher and sociologist A A Zinoviev appropriately names a community of people a *human anthill* (Zinoviev 2002). The human anthill occupies and uses a certain space (a territory), possesses relative autonomy in the internal life, makes or extracts means of existence, protects itself from external phenomena threatening its existence. The members of a human anthill have regular communications with the other members of the human anthill and live a joint historical life, that is, from generation to generation reproduce similar *individuals*[1]. There is a

[1] Let us remark that the word 'individual' stems from Latin *individuus*, that means 'indivisible'. The least indivisible particle of a society (atom of society) is the person similarly to that the least indivisible particle of a substance appears to be an atom (*ατομοσ* that is not capable to be cut, indivisible in Greek). The social organisations are analogues of molecules.

doi:10.1088/978-0-7503-3451-8ch11

division of functions among them, people occupy various positions in the human anthill, and these distinctions are partly inherited biologically, and mainly are obtained. The human anthill possesses internal identification, that is, its members recognize themselves as those, and the other members accept them as members of the community. It possesses also external identification, that is, the people who are not belonging to them, but somehow colliding with them, accept the population as an association to which they do not belong, and members of the human anthill recognize them as aliens.

Describing society, we should not forget that a set of humans is a biological population, which obeys the natural, biological laws; individuals of a human population reproduce themselves similarly to individuals of any biological population according to the programs, recorded in hereditary molecules of individuals, that are readable by means of a genetic code (see section 9.4). In this respect the human population is similar to any other biological population, but the society (for sociologists) is not only a set of individuals making it, but a complex of all of those varied relations, by which its members are connected among themselves.

In any population it is possible to notice some organization of relations of individuals, but only in the human population, at interoperability of individuals, is there a second signal system—the developed language that has appeared the greatest advantage for humans. During the mutual life of people, the rules of interaction—which are kept in the memory of people as samples of behaviour (archetypes) passed from an individual to an individual by means of oral and written texts, recorded in human language—have been developed. Owing to storing of messages and their transfer from generation to generation there is *a functional social structure*, which exists as memory about heroes and events, as production and distributive schemes, as a system of repeating samples of behaviour, fixed in habits, traditions, rules, laws and so forth. A known Russian biologist B M Mednikov has uncovered, that the texts, made in human languages, could be considered as analogues to the texts, recorded in hereditary molecules of individuals, which are readable by means of the genetic code. It has led the researcher to the assumption of a possibility to describe mechanisms of evolution of language by use of concepts from the area of evolution of genetic texts (Mednikov 1976).

The basis of the integrity of society is, of course, that each individual needs the help of others and, in pursuit of his goals, interacts with other individuals. In the functional structure of society, relatively independent stable 'branches' can be identified that regulate social relations of a certain kind. To denote this kind of branch, sociologists use the concept *institution*. The most important of the institutions is the *institution of marriage and family*, which regulates the establishment and regulation of kinship relations. Another example is the *institution of authority and government*, which regulates the hierarchy of subordination in the implementation of generally significant projects. The system of the economic relations, which will be discussed in more detail in section 11.4, is also an example of a public institution. We can say that the functional structure is a set of mutually intersecting social institutions.

In one way or another, the functional structure is legitimized, and, in fact, society as a whole is perceived as a legitimate functional structure that determines the social order in society, its integrity and stability. A society arises as a common one for heterogeneous people and their groups with different interests as a condition for satisfying their private interests, but at the same time there are projects important for the general welfare that can only be accomplished as a result of joint efforts. In the simplest case, the society is considered as a non-structural (ethnic group) and according to the definition given by Dugin (2011), 'is a group of people who speak the same language; recognize their common origin; have a complex of customs, ways of life, preserved and sanctified by tradition, distinguished from the customs of other groups'. These characteristics are inseparable from any assembly of people, which, in addition, may have a more developed social functional structure, including the institutions named above. The functional structure is stable and reproduces from the codes stored in the texts, but it is nevertheless subject to changes. Over time, there is usually a complication of the functional structure from less differentiated to more differentiated, from simple to complex. Society is a self-regulating system, and the sources of all changes should be sought within society.

11.2 Functional social structure

The functional structure of the society exists in an intangible form, in the form of rules, laws and habits. The individual lives among other individuals, being conformed with their reaction to his actions. Except for messages directed from one individual to the others, each individual in a society is under influence of messages whose origin can be unknown and insignificant for the recipient. There is a huge file of messages 'poste restante', the postponed messages (various records, books, films and other), which can be found by other individuals and are understood. The material remains of activity of humans also are messages which can 'be read through'. There is a field of messages, which have gone off the sources and exist in themselves.

11.2.1 The status of an individual and socioconfiguration

Each individual lives and acts within the constraints of legitimate functional structure, adhering to the standard system of values and participating in many, if not in every, institution; one can say that the individual occupies a certain position (has the status) in the system of relationship. The social status defines a role of an individual in the social life, as it is understood in the version fixed at the current moment by the functional structure.

First of all, each individual has characteristics that trace an individual state of the person and are not defined specifically by their social status, such as ethnic accessory, age, health, education, capacity for research or arts and so on. For example, it is possible to consider each person to be healthy, or practically healthy, or ill—in this case variable 'health' can accept three values. Their mental health or talent also can be characteristics of the individual. A M Plushchevsky (2001) describes a variable 'passionarity', going back to Gumilev (1990), and ways of its measurement: table 1 of his article defines seven values of this variable.

For the characteristic of the role of an individual in society, it is useful to introduce the variables, which values define the place that the individual occupies in functional structure, in social institutions. Such a characteristic can be, for example, 'participation in government', that is, in what quality and in what measure, the individual participates in the control system, or how much he subordinates himself and how much he has influence on others. One more example can be the amount of income of the individual for maintenance of his own demands at the expense of certain efforts. An important variable is 'wealth'—a part of national property that is at the disposal of the person, including possession of means of production.

The listed examples show that, to locate a personal state and a social status of the individual, some quantities are being introduced; these quantities are established and estimated as a result of empirical sociological research. Though we are still very distant from the knowledge of the full set of independent characteristics that should be attributed to the individual to describe his/her state and behaviour, it is possible to assume, at formalization of the description, that s independent parameters can be attributed to each individual, while each parameter accepts value in a fixed (discrete or continuous) set, so that the status of the individual in social space is specified by the vector $\boldsymbol{r}$ with s components

$$\boldsymbol{r} = \{r_1, r_2, \dots, r_s\}.$$

Each component of the vector accepts one value in the fixed set of values, so it is possible to speak about a point in the s-dimensional space that, following Weidlich (2000), we name as *social space*. Here we assume, that all the necessary-for-description variables are included in the set, though, certainly, they can be classified on that or this basis in advance.

Considering an ensemble of individuals, it is convenient to believe, that the s-dimensional social space can be divided into cells, and the state of the population can be set with the numbers of individuals in each cell of the social space $n_{r_1,r_2,\dots,r_s}$. The distribution of members of the population on alternative positions within the limits of the given fixed set defines *socioconfiguration* $\{n\}$ that, certainly, changes in due course. These statements recall the description (see chapter 8) of distribution of particles of ideal gas over cells of the phase space of a single particle[2]. In the simplest case, for example, when $s = 1$, the point in this space is characterized by a quantity that accepts scalar values, and the social configuration of the population is specified as a set of numbers: $n_1, n_2, \dots$.

For the description of the system, one can introduce a function of distribution $n(t, r_1, r_2, \dots, r_s)$, thus $n(t, r_1, r_2, \dots, r_s)\Delta r_1, \Delta r_2, \dots, \Delta r_s$ defines the number of individuals having values of variables in the element of phase volume

[2] At the construction of fundamental theory, Weidlich (2000) considers an ensemble of configurations of the system and introduces a function of probability of realization of socioeconfiguration $P(\{n\}, t)$. This approach recalls Gibbs' approach (see chapter 5), while the approximation discussed further recalls the description of ideal gas by Maxwell and Boltzmann described in chapter 8.

$\Delta r_1, \Delta r_2, \ldots, \Delta r_s$ at point in time t. It is convenient to normalize function of distribution to the total number of individuals

$$N = \int n(t, r_1, r_2, \ldots, r_s)\, dr_1 dr_2 \ldots dr_s. \qquad (11.1)$$

One can notice, that the principles of construction of the theory of sociodynamics are similar to the principles that are used at the description of dynamics of an ideal gas (see chapter 8). The roles of particles of the system are played by separate individuals who, eventually, determine the behaviour of the society. However, unlike gases, where the state of each particle in the system (in the classic case) can be completely given by the values of its coordinates and impulse, and behaviour of the system of particles at the preset forces can be completely described, it is not clear how many and what parameters one needs to locate a state of the individual, the principles of behaviour of a separate individual are not quite formalized. Nevertheless the developed concepts allow starting quantitative research of the social system. Similarly to the ideal gas consisting of atoms, we can consider a social system as an ensemble of the separate weakly cooperating individuals.

11.2.2 Strata, domains, classes and castes

The existence of the individual characteristics (variables of state of the individual) gives the basis for separating some groups of individuals within the considered community. Under integration of the function of distribution $n(t, r_1, r_2, \ldots, r_s)$ in respect to some variables, it is possible to find functions of distribution for the limited number of variables. For example, under integration over all variables, except for r_j, we find function of distribution of individuals over a variable r_j

$$n(t, r_j) = \int n(t, r_1, r_2, \ldots, r_s)\, dr_1 dr_2 \ldots dr_{j-1} dr_{j+1} \ldots dr_s. \qquad (11.2)$$

This function represents the splitting of the entire population over *strata* (layers) according to the selected characteristic.

As a stratifying index, various characteristics can be chosen, both private such as age or education, and social such as economic or political indicators. Most essential meaningful variables are established empirically and their choice depends on objectives of research; various researchers lay various variables in a basis of stratification. With various choices of variables, the strata get special names: for example, splitting of individuals on the basis of position in a legitimate state structure defines *estates* or *castes* (Kordonskiy 2008). The relationship of the people to the means of production and the character of assignment of goods defines *classes*[3].

[3] Note that the concept of the social class has appeared irrespective of concept of stratum. The known expert on Marxism, V I Lenin defined classes as follows: 'Classes are the great groups of people differing on their places in historically certain system of a social production, on their relationship (mostly fixed and issued in laws) to means of production, on their role in public organization of work, and, consequently, on ways of reception and the sizes of that fraction of public wealth, which they have'. Lenin V I, *Works*, 4th edn, vol 29, p 388.

Note that individuals are incorporated into stratum only by the presence of the corresponding index that does not assume mandatory intensive interoperability between them; nevertheless, the general meaning of this index gives a basis for interoperability. Let us notice, that in the basis of concepts of stratification of the society lay the ideas of Karl Marx and Max Weber about classes in the industrial society and social interoperability. These concepts have received further progress in Parsons's and in West-German sociologist Dahrendorf's works.

For more detailed investigation of social functionality, it is possible to use not only one, but several indexes, getting the intersection over the strata or non-functional groups which are convenient to name as *domain*. For two variables, for example, one has

$$n(t, r_i, r_j) = \int n(t, r_1, r_2, \dots, r_s)\, dr_1 \dots dr_{i-1} dr_{i+1} \dots dr_{j-1} dr_{j+1} \dots dr_s. \qquad (11.3)$$

The existence of the arbitrary great number of variables for the characteristic of an individual makes the description of social architecture very complicated; taking even two parameters, a status position of the individual cannot be defined as something unequivocal, we cannot estimate position of individuals in the terms 'above–below'. It leads to impossibility of allocation of the best people—*elite of the society* without any additional preferences that are influenced by dominating ideology or public moods. Each time appoints its own heroes (Bukharin and Malkov 2014).

Certainly, the splitting of a society according to one or even several indexes coarsens the structure of the society, but allows one to track the social changes. So, the concept of class was used by Marx and his followers for the description of the most essential processes of the capitalist society.

Example: the Soviet society before perestroika

An interesting example of stratification of society was provided by Kordonskiy (2006). The basic 'formally-registrated' characteristics of individuals in the Soviet society included the variable 'participation in the institute of government', which, according to Kordonskiy (2006, chapter 8) accepts three values: a functionary, a philistine and a dissident. 'The strata of functionaries, philistines and dissidents have been connected among themselves by the repressive and distributing mechanisms, providing the social stability and controllability by the society as a whole ⋯. The association of each person to the stratum was located by the state in obvious and implicit terms (the functionaries in the form of the nomenclature, the philistines in the form of 'ordinary humans'—inhabitants, and dissidents as dissidents) (Kordonskiy 2006, chapter 8). For a more detailed characteristic, apart from the mentioned variable, Kordonskiy considers necessity to take into account 'the form of activity' and 'dialects of language', which, in fact, can be considered as a variable 'attitude to the institute of authority'. This variable also accepts three values: 'approval of the system', 'indifference to institute of government' and 'denying of institute of government', which allows one to introduce, independently of formally-registrated stratification, another strata according to respect for the institute of authority.

Table 11.1. The domains of Soviet society before perestroika.

Measure of approval socio-registration level	Support	Indifference	Denying
Functional	E K Ligachyov	Nomenclature gossips intriguers	A N Jakovlev
Dissident	Roy Medvedev	Children of Arbat	A D Sacharov
Narrow-minded	Informers and tale-bearers	Nikolay Nikolaevich	Authors and performers of political jokes

As a result of the splitting of society over two variables with three values for each, nine domains are produced (see table 11.1). The first line of the table contains the list of the categories formally participating in the government, but differing on a level of support of this structure, while the first column shows the types of the individuals supporting the existing authority system. It is possible obviously to characterize each domain of the social structure that has been connected by function into the entire state system. The names of domains are given by Kordonskiy (2006, chapter 8) according to the names of typical representatives.

The division of the society over the domains certainly helps to describe the events of 'perestroika', that is, the destruction of statehood. Stratification of another sort, which is necessary for the description of events in an economic field of activity, is also discussed by Kordonskiy (2006, chapter 8).

11.2.3 Groups and organizations

Unlike stratum or domain, *the functional group* is an ensemble of individuals cooperating with each other on a regular basis. The individuals that are allocated within the group interact with each other more intensively, than with the others (Menger 1938, 1983). The functional groups can be united in *the organizations* that represent greater associations of the individuals operating on some coherent plan. The greatest group or organization is a society as a whole. The individual can participate in many functional groups.

The group, as the whole, is self-organized for coordination of actions of the individuals, which are directed to the achievements of specific objectives. The group declares and represents essential interests of all participating individuals, but not all participants of a group realize the existence of the general interests. In other words, if one individual does not have demand for the social good, it does not mean that society as a whole does not require it (Samuelson 1954). In such a situation, when each individual has some interests, preferences and objectives, the disputed situations, when preferences of an individual will not be coordinated with requirements of the group, are possible. The special interests of some groups or a society as a whole, which are not reduced to individual interests, can exist; such preferences do not come to light in individual preferences (Grinberg and Rubinstein 2005).

Each individual plays many roles and can belong to several groups at the same time, being, certainly, a member of a society as a whole, therefore the sum of numbers of members of all groups appears arbitrarily greater than the number of members of the society. The splitting of a society into functional groups appears, as a rule, impossible. However, such choice of an index of stratification is also possible, that every stratum appears to be a functional group. An example is the splitting of society on the basis of an association with a certain political party—a case that was considered by Weidlich (2000, chapter 5).

11.3 Dynamics of the social structure

It is impossible to describe chaotic behavior of individuals that are similar to particles in an ideal gas; instead of it, a society is perceived and depicted as a specific functional structure, in which one can identify institutes, strata, domains, groups, organizations and other components. Social evolution can be understood as processes of emerging and differentiation of social structure, as progress of the society from simple to complex forms of organization. The society is a self-adjusted system, and origins of all changes should be looked for inside the society.

11.3.1 Mechanism of changes of the system of values

The existence of a society depends on cooperation which assumes an agreement of members of a society concerning the basic values and, consequently, *the system of values*, which are considered as a main constitutive element of social life, which lays in a basis of legitimate functional structure. The system of values (the ethical system) should be recognized in social consciousness and, thus, institualized as the sample that defines the type of social system, and is defining in an establishment of tendencies of progress of a society. The system of values defines *ideology of a society*—the system of views and ideas about attitudes of people to the existing reality and to each other.

In the base of the functional structure of the society, a system of values is laid, and we should specify the mechanisms of recognition, definitions and formulations of the system of values, which cannot be external, they should exist within the society. During ancient times, it was a function of animistic philosophy that was both a science, and art, and the arch of ethical rules. In modern societies, a special institute that has a functional role to formulate and study the social values exists and is supported.

Such a social institute is art that has existed for millennia. Asking themselves the question, what is art, researchers (see, for example, Feinberg 1987, 2004) described many characteristics and features of art and guessed that art executes a very important function within the society. Not diminishing importance of the listed features of art, it is possible, referring to the analysis of the problem[4], to state, that *the basic social function of art is formation, investigation and maintenance of the system of values*. Art is a system of creation of basic latent stereotypes of behaviour and development of sense of existence of the society. It is possible to see that only art is predetermined for realization of these functions; other social institutes appear unsuitable for this role. Science, for example, cannot cope with the problem, its basic function is different: to order the experience of the humans.

In any case, the specified function of art is recognized by those who create works of art, for example, writers who are fairly named as engineers of human souls. The Russian writer Leonid Leonov, for example, in a note 'The Memorable Remark' (*Literary Newspaper*, on June, 18th, 1986, Nov 25, p 5) wrote: '⋯ the public importance of our craft is ⋯ a preliminary ⋯, long-range investigation of the future, getting into the depth and essence of the enormous and frightening uncertainty.'

A work of art often does not state a lesson explicitly, and, apparently, sometimes it is impossible to do: words, notions, concepts are still only developing and being defined. New ideas happen to be presented in a controversial form, in the form of a question pointing to a hint of the answer. Sometimes the answer arises in the reflections about behaviour of 'the wrong heroes'.

The various forms of art are important. Keller (2007), for example, writes, that music bears an especially strong 'ideological' charge. Music approves essential value: a condition of the spirit to be ready to act[5]. The impact of a drum operates, conducting behind itself a rhythm of the human heart, for anyone to whom it would belong, regardless of ethnicity. The bravura march calls on fight, and the mourning one casts grief. The song can be considered as an impact on psyche. The listener is fascinated by sounds, runs into the necessary condition and ⋯.

New values do not always win general recognition at once; it is exception as a rule. New ideas (value) are accepted by a group of adherents, functional group (a party) which can increase in numbers and capture influence of the whole society or disappear. The change of values causes evolution of functional structure, which changes through actions of separate individuals, and in this way the adaptive capacity of the society, as a rule, amplifies.

---

[4] The Russian philosopher Gorskii (1985, p 181) remarks, that the objective and purpose of art is 'to cause in the addressee aesthetic feelings and experiences, on the basis of which and in concordance with which, both the ethical and social ideals and values are realized, and educational role of a work of art is carried out.'

[5] As described by A P Platonov in 'The Foundation Pit': 'The music has approached to the barrack and has played special vital sounds, in which there was no idea, but there was the exulting presentiment resulted Vostchev's body in the jingling condition of pleasure. The disturbing sounds of sudden music gave feeling of conscience, they offered to care the life time, to pass a distance of hope up to the end and to reach it, to find there a source of this agitating singing and to not begin to cry before death from melancholy of futility'. Let us quote also the words of the hero of 'The Kreizer's Sonata' by Leo N Tolstoi: 'The music forces me to forget myself, my true position; it transfers me to any other, not my position; it seems to me, under influence of the music, that I feel what I, actually, do not feel; that I understand what I do not understand; that I can what I cannot'.

11.3.2 Mechanism of changes of the functional structure

The number of the status variables $r_1, r_2, \ldots, r_s$ contains both the characteristics of the individual, and their position in strata and social organizations. The function of density of distribution of individuals $n(t, r_1, r_2, \ldots, r_s)$ represents existing functional structure (the institutes, strata, groups, organizations and other components); the quantity of any component can be found from the function of distribution with the use of suitable operation (see, for example, formulas (11.2) and (11.3)). The function of density of distribution of individuals, generally speaking, changes at transitions of individuals from one state to another, so that the social changes can be described as changes of this function. The change of state of an individual is defined by both decisions of the individuals and independent-of-him circumstances. Among the last causes, one can discover not only the changes connected with natural processes in the organism, but also all conditions of the life of the individual: the circumstances of dwelling, the encircling of the individual by other people and so on.

In the case when the conditions of existence of individuals remain constant for a very long time, it is possible to think that at mutual influence of variations of functional structure and behaviour of individuals, the society comes to some steady state, at which the functional structure does not change. Thus, in this situation

$$\frac{\mathrm{d}n(t, r_1, r_2, \ldots, r_s)}{\mathrm{d}t} = 0. \tag{11.4}$$

However, every individual in this situation passes naturally from one state to another: the organism grows, gets older and dies; there is an accumulation of knowledge, acquisition of qualifications, experience, which allows the individual to pass to a higher social status in the constant functional structure of the society. It is possible to believe that the society represents some invariant structure despite the natural variations in status of individuals, such as changes of age, education, career and so forth.

The condition of steady state is the strict obedience by each individual to written and unwritten rules, dictated by existing functional structure. However, the behaviour of individuals appears not always to be *conformed.* The individuals can make mistakes in their activity, make actions, unforeseen by their roles in the functional structure, and even consciously aspire to make changes of the structure. The behaviour of the individual is determined not only by his position (status) in functional structure, but also by his opinion of the efficiency of the existing structure and his position in it. Naturally, deviations from the prescribed behaviour cause public misunderstanding, reproach and punishment. All social norms are accompanied by the sanctions protecting from non-conformity. The society has special institutes in the form of police, courts, prisons etc for maintenance of legitimate functional structure.

Considering social changes, it is possible to neglect accidental deviations in behaviour of individuals from the roles prescribed by their status and consider only regular deviations leading to the intense tensions between individuals and legitimate functional structure. If disputed situations are resolved peacefully, by compromises and reforms, the legitimate structure evolves. If the functional structure is fixed rigidly and does not support change, there can appear functional parallel structures, shadow structures. In this case, disputed situations lead to significant (revolutionary) changes of the legitimate functional structure.

11.3.3 Kinetic equation

To find the equation for change of function of the density of distribution of individuals over status variables $n(r_1, r_2, \dots, r_s, t)$, we introduce the probability of the changes in position of the individual on switching from one state to another in the unit of time $w(\boldsymbol{r}', \boldsymbol{r})$, where $r_1, r_2, \dots, r_s$ and $r_1', r_2', \dots, r_s'$ are the initial and final states of individual. The function of transit of the only individual $w(\boldsymbol{r}', \boldsymbol{r})$, depends, generally speaking, on all other individuals. To get a realistic account of the problem, it is necessary to introduce some assumptions at this stage. It is possible to admit that influence of individuals on the rules of transition is described by some average quantities, which are characteristics of functional structure. Then, the decrease of the value of the distribution function in a point $r_1, r_2, \dots, r_s$ can be recorded in the form

$$\sum_{r_1', r_2', \dots, r_s'} w(r_1', r_2', \dots, r_s'; r_1, r_2, \dots, r_s)\, n(r_1, r_2, \dots, r_s, t).$$

Similarly, considering all possible transitions from any states of individuals $r'_1, r'_2, \dots, r'_s$ to the given state $r_1, r_2, \dots, r_s$, we find the increase of value of the distribution function in the point of social space that is interesting for us

$$\sum_{r_1', r_2', \dots, r_s'} w(r_1, r_2, \dots, r_s; r_1', r_2', \dots, r_s')\, n(r_1', r_2', \dots, r_s', t).$$

Thus, we can record the kinetic equation for the distribution function in the simplest approximation

$$\begin{aligned}\frac{\mathrm{d}n(r_1, r_2, \dots, r_s, t)}{\mathrm{d}t} = \sum_{r_1', r_2', \dots, r_s'} &[w(r_1, r_2, \dots, r_s; r_1', r_2', \dots, r_s')\\ &n(r_1', r_2', \dots, r_s', t)\\ &- w(r_1', r_2', \dots, r_s'; r_1, r_2, \dots, r_s)\\ &n(r_1, r_2, \dots, r_s, t)].\end{aligned} \tag{11.5}$$

Let us remark that the recorded equation coincides with the equation (3.58) for average values of numbers of individuals from the third chapter of the book by Weidlich (2000).

It has been assumed above that influence of the environment on probabilities of transition $w(r_1', r_2', \dots, r_s'; r_1, r_2, \dots, r_s)$ is taken into account by some collective variables describing existing functional structure. But it could happen, certainly, that transition of an individual to a new condition can be specifically connected with states of one or several other individuals. In this case, when the transition is determined, for example, by the 'collision' of two individuals, it is necessary to record the two-individual terms in the right part of equation (11.5). Various approximations could be considered.

11.3.4 Rules of change of individual status

Among different desires, by which individuals are being guided, one can find, first of all, the simple desires to find comfortable shelter and good food. The individual chooses aims and ways of actions for achievement of the chosen objectives within the limits of existing functional structure. The behaviour of the individual is shaped also under the signals received from other individuals. Thus, the individuals are partially free in their decisions, and partially depend on the system that encourages acceptance of this or that decision.

In the 'non-collisional' approximation, which was assumed at the record of the kinetic equation (11.5) in the previous section, the changes of the distribution function are connected with probability of transition $w(\boldsymbol{r}^{\gamma}, \boldsymbol{r}^{\alpha})$ of the individual from the social state $\boldsymbol{r}^{\alpha}$ to the state $\boldsymbol{r}^{\gamma}$ in the unit of time. It is possible to state some motivation, allowing one to reduce the set of probabilities of transition to a universal function of variables of the only individual. It is supposed that there is a function $u(\boldsymbol{r}^{\gamma})$, such that

$$w(\boldsymbol{r}^{\gamma}, \boldsymbol{r}^{\alpha}) = w_0(\boldsymbol{r}^{\alpha}) \exp[u(\boldsymbol{r}^{\gamma}) - u(\boldsymbol{r}^{\alpha})]. \tag{11.6}$$

The function $u(\boldsymbol{r}^{\gamma})$ depends on the variables of state of the only individual, and can be referred to as function of utility or, according to Weidlich (2000, chapter 3), as to motivation potential. In considered approximation, the motivation potential depends only on variables of the individual, to whom it relates, whereas the potential includes also the characteristics of the functional system. The internal tendencies can be described as the tendencies leading to maximization of the potential of satisfaction that are a function of states of the person. The behaviour of the separate individual is being described basically in correspondence with the requirements of food, a refuge and reproduction, and the individual chooses their social position according to internal tendencies and external restrictions (influences). Optimization of the utility function at certain restrictions defines a preferential state of an individual, which on comparison with an actual state allows one to estimate a measure of internal strains[6]. There are also other ways of describing preferences of individual.

In the 'non-collisional' approximation, the situation can be described by means of unique function of utility, but in 'collisional' approximations, when mutual influences of two or more individuals are considered, it is necessary to use two or more functions of utility, every one for a separate individual. In this case, we should optimize the two functions simultaneously. Generally it is possible to consider a problem of behaviour of the all individuals, optimizing the set of functions of utility for all individuals. To the set of individual utility functions, we have to add the

[6] As an example, the choice of a set of products for consumption can be indicated.

utility functions for the system, as a whole, and such reasoning leads to the many-criterion problems that were introduced into consideration by Pareto (1897). For the formulation of the problem, it is necessary to use all our knowledge and additional assumptions about the behaviour of people in various situations.

11.3.5 Mechanism of decision-making

From time to time, the individual, under the given conditions, is urged to make a decision on changes of their status, being based on the received information and their own inwardness. Sometimes, the problem can be reduced to a problem of a choice of the only realization from a possible set. It is possible to tell that an elementary act of behaviour is a procedure of a choice of one variant in many. And in this respect, the study of the mechanism of execution of the choice lays in the base of researches on behaviour of a person. The research can begin with the elementary situation, when a person is asked to sort absolutely identical objects, for example, pairs of beans or balls into good and bad (see Lefebvre 1990). The tests show paradoxical results: despite special precautions to guarantee the similarity of objects (balls), results of experiment show constant asymmetry: 62% of objects appear 'good' or 'bad'. The result is explained by the statement that there is a reflective mechanism of the assessment of decision, whether the ball is 'good' or 'bad' (Lefebvre 1990). Further, we follow the description of the mechanism of assessment proposed by Oleg Kalyuzhny (2005), who has given the elegant solution of the problem about decision-making.

Let us consider the process of separating a set of identical balls into two groups: the 'good' balls are separated in group A, the 'bad' ones—in group B. Considering each subsequent ball, the investigator, with probability 1/2, makes a decision that a ball is 'good' (i.e. puts it in group A); or, also with probability 1/2, does not make such a decision and the ball becomes a candidate to move into group B. Then, one can make (with the probability p) a decision that a non-accepted ball has to go in group B or (with the probability $1 - p$) that the ball does not go, and it remains for further consideration. Reflections on the next ball begin with the very beginning. Briefly, according to Kalyuzhny (2005), these non-conceit reflections can be described as follows: And whether this ball is 'good'? (yes—A). If it is not 'good', whether my idea about the ball being not 'good' is good? (yes—B), if it is not, one has to begin all over again, that is: And whether this ball is 'good'?$\cdots$ and so on. This reasoning allows one to estimate a fraction of 'good' balls

$$P(A) = \frac{1}{2}\left[1 + \left(\frac{1-p}{2}\right) + \left(\frac{1-p}{2}\right)^2 + \cdots\right].$$

Here $\left(\frac{1-p}{2}\right)^k$ is the probability of the situation, when the ball goes in group B k times, but the decision has been recognized as incorrect. We shall note also, that, at a unconscious tendency towards 'good' balls, the probability of getting a ball in group A is more than 1/2, since the ball can get in it not only on the first, but also on any other step.

Folding the series recorded above, we shall obtain an assessment of fractions of 'good' and 'bad' balls:

$$P(A) = \frac{1}{1+p}, \quad P(B) = 1 - N(A) = \frac{p}{1+p}. \tag{11.7}$$

The fraction of the 'bad' balls in the group $P(B)$ can be identified with probability of getting a ball in the 'bad' group $1 - p$, and, thus, we find the equation for probability:

$$p^2 + p - 1 = 0. \tag{11.8}$$

This relation is the equation of the gold proportion with the unique positive decision $p = 0.62$. Thus, the fraction of 'good' balls, according to parity (11.8) is equal 0.62.

The other way describing the process of the decision-making was proposed by Lefebvre (1990). To describe it, we shall introduce some variable as characteristics of the individual. Let $I = a_0$ be a variable 'intention to say' with values 'yes' (1) or 'no' (0). In a case when there is no preferential situation, the variable takes value 0 or 1 accidentally. Then, the individual has an internal idea about his intention, and the assessment of the chosen intention (1 or 0) can be also positive (1) or negative (0). Let $R_1 = a_1$ be a variable describing the assessment of intention of the person to say 'yes' (1) or 'no' (0). It is possible to assume further that this assessment can also be estimated, so that it is necessary to introduce one more variable, let $R_2 = a_2$, describing the assessment of the previous assessment of the intention of the person to say 'yes' (1) or 'no' (0). The process of the introduction of variables in this manner can be continued, so that the individual can be characterized by values of variables

$$I, \; R_1, \; R_2, \; R_3, \; \ldots$$

In the situations when there is no preference, each variable accepts values 0 or 1 accidentally.

It is possible to assume, that the final assessment of the situation by the individual is a function of the listed variables

$$U = U(I, \; R_1, \; R_2, \; R_3, \; \ldots).$$

Lefebvre (1990) has specified this function as a power function

$$U = I^{R_1^{R_2^{R_3 \cdots}}}. \tag{11.9}$$

Table 11.2 contains the calculated values of function U for a case of two levels of the reflection for possible values of variables. In the situations, when there is no preference, each variable accepts values 0 or 1 accidentally, so that it is possible to estimate average value of the quantity $\bar{U}$, which is named the status function. At levels of reflection: 1, 2, 3, 4 formula (11.9) defines appropriating values of the status: 0.75, 0.625, 0.312 5, 0.6875. The empirical value of the status $\bar{U} = 0.62$ is conformed with the value calculated for two levels of the reflection.

Thus, it is shown, that the non-conscious assessment of intentions (the reflection) appears important for the process of decision-making, whereas the reflections of the

Table 11.2. Estimation of the status function.

I	1	1	1	1	0	0	0	0
R_1	1	0	1	0	1	0	1	0
R_2	1	1	0	0	1	1	0	0
U	1	1	1	1	0	1	0	0

first and second levels appear, apparently, especially essential. The mechanism of decision-making appears universal for all people, it depends on neither age, nor gender.

11.3.6 Formation of collective opinion

Following Weidlich (1994, 2000, chapter 6), we shall consider a population consisting of $2N$ active individuals, who should choose some social rule. Each individual should show one of the two opinions: to agree (+) or to disagree (−) with the proposed rule. It is convenient to introduce variables that designate the number of positive and negative answers, as well as a deviation of numbers from average value

$$n^{+} = N + n, \qquad n^{-} = N - n. \tag{11.10}$$

Besides, we have to take into account that each individual reflects and estimates their answer, which can be formalized by introduction of the second variable accepting two values. The individual can be definitely convinced that his/her answer, whatever it is: positive or negative, is proper and is not subject to variation (the variable accepts value +), or that the individual has some worries and doubts in correctness of the statement (the variable accepts value −). Thus, for the description of the situation, the whole set of the individuals is broken into four non-functional groups (the domains)

$$n_{+}^{+}, \quad n_{-}^{+}, \quad n_{+}^{-}, \quad n_{-}^{-}$$

that have the obvious property

$$N = \frac{1}{2}\left(n_{+}^{+} + n_{-}^{+} + n_{+}^{-} + n_{-}^{-}\right). \tag{11.11}$$

The top indices designate the explicit demonstration of the choice made by an individual, while the bottom indices designate approval (+) or disapproval (−) of the preference shown by the individual.

We assume, that the individual can change his state, altering a value of only one of the variables. From a state of doubt $[^{+}_{-}]$, for example, he can pass to the state of certainty $[^{+}_{+}]$, or other state of doubt $[^{-}_{-}]$, and from the state of certainty, he can pass only to the state of doubt; the transitions can be represented schematically as

$$
\begin{array}{ccc}
\left[{}^{+}_{+}\right] & \leftrightarrow & \left[{}^{-}_{+}\right] \\
\updownarrow & & \updownarrow \\
\left[{}^{+}_{-}\right] & \leftrightarrow & \left[{}^{-}_{-}\right]
\end{array}
\tag{11.12}
$$

Distribution of individuals on the states depends, apparently, on probabilities of transition between states that, in turn, are determined by both the internal reasons and external circumstances. To describe observable preference of this or that answer, it is necessary to introduce various probabilities of transitions and to define their possible dependence on variables of the problem. In the simplest case, when all probabilities of transition are identical, the state numbers, irrespective of initial distribution, appear, eventually, in steady state and to be equal.

Considering one of the elementary versions, we shall designate probabilities of transitions between states of doubts in directions of increase in numbers of positive and negative states, p_+ and p_-, accordingly, and probabilities of all other transitions p, so that the kinetic equations can be recorded in the form

$$
\begin{aligned}
\frac{\mathrm{d}(n_+^+)}{\mathrm{d}t} &= -pn_+^+ + pn_-^+, \\
\frac{\mathrm{d}(n_-^+)}{\mathrm{d}t} &= -pn_-^+ - p_-n_-^+ + pn_+^+ + p_+n_-^-, \\
\frac{\mathrm{d}(n_-^-)}{\mathrm{d}t} &= -pn_-^- - p_+n_-^- + p_-n_-^+ + pn_+^-, \\
\frac{\mathrm{d}(n_+^-)}{\mathrm{d}t} &= -pn_+^- + pn_-^-.
\end{aligned}
\tag{11.13}
$$

In the stationary case, these equations define a parity between numbers of the individuals showing 'yes' or 'no'

$$
n^+ = \frac{p_+}{p_-}n^-. \tag{11.14}
$$

The kinetic equations of the type (11.13) can be recorded at various assumptions about probabilities of transition that allow one to investigate various situations. Weidlich (1994, 2000, chapter 6) has shown that such a model, under some assumptions of dependence of probabilities of transition on the considered variables, can lead to the sharp changes of public opinion with positive to negative or vice versa, which reminds us of the description of phase transition in physics.

11.4 Institute of economic activity

In the foundation of emergence and development of the system of social production, one can find persons, aspiring to make their life more comfortable. Every person participates in economic activity as an economic subject (agent) and operates as *a producer* and (or) as *a consumer*. In modern societies production activity is organized in scales of the entire society, and each economically active individual appears to be a member of this or that group (an enterprise, an organization, a firm). There are contradictions between social production and personal consumption, between discipline of directive planning and personal free entrepreneurship. These contradictions exist objectively, and the main problem that was thought of by outstanding thinkers is to find the best possible organization of production and distribution of the social product, taking into account interests of both the entire society as a whole, and the individual participants of economic processes: producers and consumers.

11.4.1 The scheme of production process

The production of things, described as a process of transformation of matter in section 8.4.2, is a core of economic activity that involves all *economically active people*. This activity brings into existence a great amount of artificial things of different kinds: constructions, various possible machines, works of art, principles of organization, results of scientific research and many others, which, in any way, appear useful to people. The value of things created by a production system in unit of time Y (gross domestic product, if one considers a nation) is completely determined by effective utilization of the available *social resources* (which are efforts of humans, possibility to use energy in the production processes and also raw materials from the environment) and can be presented in the following form

$$Y = \beta L + \gamma P. \tag{11.15}$$

This relation between output of production system Y and production factors L and P appears to be a specific form of *the main economic law*[7]. This relationship is determined by the existing state of the production system, but not any wishes of people. The marginal productivities β and γ are, generally speaking, functions of the ratio P/L; the form of the functions depends on the choice of monetary unit that measures output Y (see section 8.4.2).

The organization of production assumes coordinated activity of huge groups of people, so that modern (however, since prehistoric times) production has social character. The modern production system consists of a set of a huge number of hierarchically organized projects, in which the human being plays various roles. First of all, the human being appears an originator and a manager of each possible production project. In the simplest case, a single individual appears to be an initiator, coordinator, executor and beneficiary of the project. When more complex projects are organized, the various functions are carried out, generally speaking, by various individuals; there appears to be some *social production relationship* of the people. There exists some distribution of people according to their roles in production; it is convenient for the analysis to break all participants of production into two groups of people: business people (employers) and executors (hired workers). The relationship of these two groups (the classes) was the object of special interest of Karl Marx and his followers. Certainly, now we refuse to accept the

[7] For those, who adhere to thinking about capital K as about a production factor, I should like to remind them that substitutive work P appears to be a capital service, and there is a correspondence between the different approaches (Pokrovskii 2018, chapter 1).

extreme concepts about business people (employers–capitalists) as parasites on a body of workers and should consider, following Schumpeter (1911), their key role in production, as organizers and supervisors of the production processes.

An employer hires workers (executors) for a certain time (per hour, per day, per year). The worker belongs to the employer only per a certain operating time; during the rest of the time the person is left to himself. He can eat, spend time with family, take care of health, both his and his relatives. The situation can be described in such a manner that the employer rents an executor for a certain time for a certain payment, and nothing else interests him (Ellerman 2015). The hired workers carry out the duties ascribed by instructions, and, for his action, the employer allocates a fraction of the created product to the workers in the form of wages or salaries.

The business person (employer), as the organizer of the entire enterprise, gets (or leases) production equipment—a set of appliances, which allows attracting workers' efforts and substitutive work, which are true sources of value (see section 8.4.2), into the production process. The installed equipment appears as a result of efforts of many generations of people and its property to attract energy is a public property, though some items are using by employers, who are considered as proprietors of the production equipment. In fact, it is necessary to consider, that the production equipment is in everlasting rent by the business person-employer, but for the equipment to run, the business person-employer must attract and pay off social resources: workers and substitutive work, to say nothing of raw materials.

A money assessment of the production equipment is known as *basic production capital*, and the property of the equipment to create products, that is to increase value due to the workers' efforts and substitutive work, is attributed to the capital. However not the capital, but the capability of the production equipment to use working force and to replace efforts of humans with natural energy in the production of services and things, is the real cause of the increasing value. This ability is a main property of fixed production capital.

In widespread understanding, capital is everything that provides profit. If one has production shares, he gets dividends, if one's money lies in the bank, he receives a percentage as interest. The shares and money are capital in the wider understanding. However, money and the shares are only symbols that bring nothing without huge work on production of value within the framework of the capitalist organization of the national economy. The mystical force of the capital to bring profit results from the rules of distribution of the social product created by workers and substitutive work. Only efforts of humans (in view of a substitution effect as explained in section 8.4.2) lead to increase in the value of the capital.

There are different approaches to description of activity on creation and distribution of products; a heuristic model can be chosen, for example, as an ensemble of producers and consumers (Walras 1874), or as a set of interacting sectors (Leontief 1986). A review of some of the phenomenological results from a physicist's point of view can be found in the book (Pokrovskii 2018).

11.4.2 Problem of distribution of the social product

Apart from the economic laws, an example of which is relation (11.15) between output and the used social resources, the economic life of the society is determined by rules that are established by people, in particular, by *rules of distribution* of the created social product. These rules are formulated by people in favour of all or some notable people. The formulated rules, then, are reformulated in laws, declared by parliament, recorded in the constitution: rules of distribution of a social product are legitimized, and the life of the society (the social relations, including the production ones) are organizing in such a way to guarantee obedience to the recorded rules. The diversity of social systems is essentially connected with rules of distribution of the social product. The capitalist system is based on the rules of distribution, according to which the created social product is considered to be a property of owners of 'capital' that is understood as both direct means of production (the physical capital), and various certifications on the property rights (money, shares and other papers). In contrast to the capitalistic organization, alternative approaches assume that the surplus product is a national property.

Surplus product

The set of created products—the output of production activity Y is measured by monetary units and appears a function of production factors: labour L and substitutive work P (see equation (10.15)) that are considered as the consumed amounts of the social resources: workforce and the productive energy. Naturally, expenditures on the restoration of the social resources should be compensated from the output, if, certainly, the organizers of production do not set up to corrupt the social production system. Apart from payment for consumed intermediate products, the business person pays operational expenses for the production factors: in the form of wages and salaries w—to the hired workers and compensation of cost of the depleted equipment as a payment for use of substitutive work $p = \mu K/P$. When the current production expenses are subtracted from the social product Y, there still remains some quantity named (according to Marx) surplus product

$$\Delta Y = (\beta - w)L + (\gamma - p)P. \tag{11.16}$$

The proprietor–tenant of the production equipment is traditionally considered as owner of the surplus product, and it is necessary to agree with this, as the business person is an organizer and coordinator of production processes, but under one important clause: the used public resources should be completely compensated. It appears that there is an agreement between business person and a hired worker about the amount of payment, w. However, the social mechanisms of urging workers to accept a lower level of wages and salaries have arisen, which allows the business person to maximize profit and, accordingly, to minimize the expenses, so that the payment to a working person w for the specifically performed work, as a rule, does not compensate all expenses (education, healthcare $\cdots$) on restoration of the human population and even, as it is described, the workforce. In a similar way, the operational expenditure of the business person on maintenance of the equipment

p does not compensate the social expenses on research and design of the appliances which allow usage of energy in creative aims[8]. Nobody can say what are the 'right' values of w and p, so that social control is needed for the society not to be corrupted.

The government has appeared to be a guarantor of maintenance and enhancement of social resources and a regulator of social relations. The main duty of the center in an economic area is maintenance of social welfare and control of economic processes to create favorable conditions for activity and development. The sovereign government (state) appears to be an economic agent, which is established to represent the general interests (the internal and external security, infrastructure, social programs, the environment and so on) of all members of the society. Thus, care of some set of social needs is compelled by the central body (the government, the state) to be taken up, because it does not correspond to the power of separate business people or because it does not bring them any immediate income. The investment in demographic and research projects, for example, give effects over years and, as a rule, do not attract private business. The more complete set of government's supervision includes power supply, transport, communication, protection against epidemics and natural disasters, roads, post service, education, information service, internal and external security, social insurance, care of the elderly and infirm, as well as care for the environment. The central body activates and performs the national projects and, for this purpose, in this or that way, the fraction of the social product created by the production enterprises should be separated. Besides, the central body is compelled to interfere in relationships between people on the basis of vague and not always adequate concepts about justice. The center accepts the duties and assigns the appropriate rights.

Payment for social resources

Thus, in the most elementary approximation, it is necessary to consider the production–distributive relations between the three aggregate economic subjects: group of business people (employers), group of hired workers (executors) and the central body (government) that, on definition, represent the national interest. Each of the economic subjects takes part in the social production and aspires to increase their share of the limited social product. Economic subjects are compelled to cooperate with each other to establish the to-all-acceptable distribution. However, a method of optimal distribution of the social product could be proposed.

The main economic law (11.15) asserts that only workers' efforts and substitutive work are true sources of value. To guaranty the compensation and possible enhancement of the used social resources, it would be fair to establish the additional deductions from the surplus product (11.16), according to the use of workers' efforts and substitutive work, and, thus, the distribution of the social product between the groups business people, hired workers and the government, accordingly, should look like

[8] Such expenses are compensated partially by payment for use of inventions and know-how in production through the various patent and license schemes.

$$
\begin{aligned}
\Pi &= (\beta - w - d_L)L + (\gamma - p - d_P)P, \\
N &= wL, \\
G &= d_L L + d_P P,
\end{aligned}
\tag{11.17}
$$

where d_L is a norm of payment for use of workforce, d_P is a norm of payment for use of substitutive work. The quantities d_L and d_P are limited: on the one side, the income of business people Π should be positive in a reasonable amount, so the norms cannot be too great; on the other side, payments should compensate an exhaustion of social resources and, if possible, provide their development, so that the quantities d_L and d_P cannot be too small.

The relationship between the norms of payment per units of expenditures of labour and substitutive work d_L and d_P are being established according to average values of the quantities: the marginal productivities β and γ, the direct compensation for the factors of production w and p, and the fraction of the income of business people in the gross final output $\pi = \Pi / Y$, and are calculated according to the relations

$$
d_L + w = \beta(1 - \pi), \quad d_P + p = \gamma(1 - \pi). \tag{11.18}
$$

The mechanism of extraction of payments for the national aims can be different, though such an allocation exists in every society. From conventional tradition, it takes a form of taxes, which represent (to say nothing of the details) deductions from profit of enterprises and personal income. However, this conventional scheme of taxation does not look fair and causes some questions: why does the money that I have earned by fair work, belong not only to me, but also to that person who did nothing. The progressive scale of taxes is unfair: a person, who earns more, has to pay more. Business people try to avoid taxes: there are some offshore schemes, the shadow enterprises, black salaries and other inventions. The right to withdraw a part of the product by the government is challenged; business people are convinced that the profit belongs to them fairly. The existing mechanism of distribution of the social product appears imperfect, though, in fact, nobody can reject that the surplus product belongs not only to the proprietor–tenant of the means of production, but also to the entire society.

Is it not better to introduce directly the payments for the use of workers' efforts and substitutive work, as shown in the distribution (11.17)? There should be also added the important payment for the production usage of the natural environment, as one of the major social resources. Thus, the three basic payments for the usage of the social resources could be established. These payments do not exclude the existence of any minor payments and fees, for the production system to be more precisely adjusted.

The advantage of the proposed principles of payment for social resources is, first, the fairness of distribution of payments on the enterprises, secondly, the simplification of the supervision of gathering of payments into the social fund (so it is easier to control the used resources rather than profit and income), and, the last, though, perhaps, the most important, such a system includes the automatic mechanism of revealing and excluding inefficient proprietors, which will lead to an increase of the labor productivity.

Efficiency of the enterprise and the entrepreneur's premium
According to Schumpeter (1911), the entrepreneur is the central figure of the production system; he is an organizer of the production project and is looking ahead to profit. His expectations are based on the rights of ownership of the means of production, which, as the entrepreneur believes, are unconditionally followed by the rights of ownership of the created product[9]. Though this is a widely spread conviction[10], one has to note, that the right of property on the created product has risen only in the case where the entrepreneur has completely paid for the used social resources[11]. After that, the created product can be considered as the property of the business person, it can be consumed, sold, destroyed, used in production, made into anything that the business person wishes, while it is supposed that he *completely* compensates the used social resources. It is one of the government's duties to control the rights of property.

The norm of necessary payment for social resources (the norm of subtraction from the product created in the production processes) can be calculated according to the rules (11.18), if value of the quantity $\pi = \Pi / Y$ is preset. The fixed value of this quantity allows estimating a middle income of an entrepreneur, if one knows the total number of entrepreneurs, and, on the comparison with the middle income of a hired worker, allows us to estimate the norm of exploitation of the hired workers by business people.

There appears to be disclosed a wonderful property of distribution (11.17) to stimulate increase of efficiency of the use of social resources. The matter is that the marginal productivities of a separate enterprise, $\tilde{\beta}$ and $\tilde{\gamma}$, differ, certainly, from their average values over the entire system. The local values of marginal productivities can exceed their average values, if the owner of the enterprise starts to issue a new product or to use a new technology, and, in this case, the entrepreneur, who, according to Schumpeter (1911), can be called a neo-entrepreneur, receives additional profit. The talent of the business person is uncovered in an effective utilization of social resources; the most successful business people deserve to get additional profit—*the entrepreneur premium*, and the enterprise has the advantage of surviving.

[9] During the centuries, this statement has seemed to be fundamental and solid, which is why the question about the ownership of the means of production appeared to be the main issue of all social transformations. In 1917 in Russia, the Bolsheviks struggled for the means of production to be nationalized; the meaning of transformation of the ninth decade of the last century in Russia—the struggle for transfer of the means of production to private hands. As in one, so in the other case, these struggles were struggles for the right to exploit the hired workers; the change of the social situation has not led to stable social relations, and it forces us to give a thought on the reasons for the existing problem. It is possible to think that the issue is not in that people own production activities, but in that they use the social resources, not caring about their maintenance. The principle of assignment of surplus value without any constraining forces destroys any society. This is a problem for each national society; as an example, it happens in Russia.

[10] David Ellerman (2007) names as 'a fundamental myth' the widespread belief, that the proprietor of the enterprise has the right to appropriate the product created by this enterprise.

[11] Apparently, this statement must be included in the law about right of property over the created product.

Others ought to be satisfied with average value of profit. In the case when local values of marginal productivities appears to be less than their average values for the entire system, the owner of the enterprise will feel difficulties and should take measures to improve the production processes, for example, to replace manual work by mechanical (this increases in the ratio P/L and, in turn, increases the marginal productivities, as is seen in section 8.4.2). Otherwise, the enterprise will be eliminated. In this way, distribution (11.17) without any additional slogans stimulates an increase of efficiency of the use of the social resources. Let us note that conventional taxation does not possess such a property. Moreover, conventional taxation oppresses production activity and stimulates concealment of profits and incomes.

By accepting the new principles of collection into the social fund, the government receives the tool for the assessment of efficiency of use of social resources, which will allow supervision of (to operate) growth of labor productivity and regulation of the distribution of social product. New rules are directed to stimulation of effective development, to creation, not to destruction. The growth of labor productivity leads to improvement of well-being of everybody, which is smoothing of contradictions and softens customs. Note that the rules can work effectively only with freedom of business, when each member of the society can organize a business according to their own wishes and initiatives. Any difficulties at registration of the enterprise should not be provided.

11.4.3 Stratification on the wealth

Adam Smith in his famous work (Smith 1976, chapter 5) wrote: 'Every man is rich or poor according to the degree in which he can afford to enjoy the necessaries, conveniences, and amusements of human life'. The activity of a human concerning finding and creation of the goods for subsistence determines the style of life that is different in various epochs. The economic social state of an individual in modern society can be determined with some basic quantities (wealth, wages, profit, consumption, debts and taxes). Some other variables, such as education, career, health and so on depend on the basic economic variables of the individual; it is interesting to find out interdependence of variables of the state, but in this section only the characteristics concerning the economic status of an individual are considered.

What is wealth?

In the sensate form, the stored wealth represents, first of all, the set of things, construction, every possible car, works of art, principles of the organization, results of scientific research etc, that appear useful to people. A source of wealth is the production activity of humans (see section 8.4.2).

Over the centuries, a system of direct estimation of value of the social wealth based on the account of the right of possession has been developed and continues to be improved. Each object has the proprietor, about which there is a record or a certificate. Some property is intended for direct personal or public use (houses, construction, cars, yachts and so forth), and some property is connected with production processes (means of production, real estate, businesses). Value of the means of production is estimated according to securities being in circulation (actions, bonds). All assessments are established, certainly, in market prices with conditional monetary units.

Direct assessments (Shorrocks *et al* 2010, table 2.4) of the world wealth for 2010 give a quantity of about 200 billion (2×10^{14}) US dollars, and about half of this amount made not financial, that is, being in private use, wealth. By other direct assessments (Brandmeir *et al* 2011) value of world financial property (world financial assets) in 2010 made 95 264 trillion of euro or nearly 1×10^{14} euro. The financial property includes commercial papers, insurance funds, bank notes, bonds, depositary certificates, checks, etc. Besides, the structure of financial property includes derivative securities, which are the financial contracts, entitling (obligation) one to buy (to sell) the certain quantity of a base active (for example, actions, bonds, oil, etc) during the stipulated timeframe under the prearranged price. Their cost, as well as the possibility of gaining income from them specifically depend on movement of the prices for other securities or commodity actives.

The resulted figures testify to the huge accumulation of the wealth of all kinds (material and non-material, personal and industrial) by humans. Further on in this section, we consider a situation from the point of view of an individual, who participates in economic activities as the subject (agent) and operates as *a producer* and(or) as *a consumer*.

Individual in the system of production–distribution

Though we try to describe the society as a whole, in macroscopic terms, it is impossible to forget that the considered object is eventually an ensemble of individuals, and the problems of connection of the variation of the social structure with behavior of separate individuals remains. An individual lives in a society, interacting with other humans. Every society has a government that, along with individual economic subjects, takes part in the organization and functioning of production. The government collects taxes and profit from the state enterprises and determines investments into the sectors of economy, according to the strategic interests of the whole society. We will proceed from a simple scheme in which the social position of a person and his opportunities are determined by the amount of personal property on disposal of an individual: personal property a and production funds (capital) k. The individual has income from his efforts in production w (wages), in addition, capital brings him an income rk; the estimated value of $w + rk$ is used for current consumption c, possible debt service qb, and other necessary payments, so that the change in personal wealth can be written as

$$\frac{d(k + a - b)}{dt} = w + rk - qb - c - \tau - \mu(k + a). \tag{11.19}$$

In this balance equation, τ is the payment of taxes to the government, c is a totally consumed product, $\mu(k + a)$ represents the value of wealth, disappearing per unit of time for one reason or another (wear and aging), μ is the coefficient of disappearing. Equation (11.19) is valid, in its general form, for any economically efficient person in any public organization.

The wealth is preserved and multiplied, when the individual gains income $w + rk$, which can be understood as income from property: each person is the owner of his ability to work, perform actions, for which he receives compensation w; ownership and management of the means of production are compensated by profit rk. Economic actors, both producers and consumers, take their actions based on their own interests and reasons, and decide how much of their income should be used and how much is saved for the future. With reasonable planning by each person, future production provides for future needs.

The government (state) affects significantly the economic situation of a person by collecting contributions to the public fund (taxes τ) and providing services (internal and external security, infrastructure, social programs, the environment, and so on) to all members of society (see section 11.4.2). Just as with individual economic entities, the government participates in the organization and functioning of the production, that is, it is an economic entity, and the question of what should be a centralized share in economic life appears often the subject of discussion. The concentration of the means of production in the hands of the government allows the central government, in the name of all society and in the interests of all people, to decide which part of the public product should be set aside for future production and which part can be used for direct consumption. In another highly idealized case of the organization of society, when all means of production are distributed among different owners, the role of the central organ in economic life can be neglected. This case of a completely disconnected market economy was studied by Walras (1874): the interests of producers and consumers are fully taken into account, but the possible interests and demands of the entire society are completely ignored. The two extreme cases illustrate two approaches to the organization of social production. Obviously, extreme cases are unstable, and the desire to implement this or that case in pure form is absurd. The social production system is an entire organism and central planned management of economic activity is necessary in one way or another.

Characteristics of non-equality

The position of an individual in a society is determined essentially by the amount of wealth $a + k$ that is supervised by the individual. This asset, eventually, brings the income x, received by the individual in a unit of time. By virtue of a balance parity (11.19), the distribution of individuals on wealth is connected with the distribution on income x, and we shall limit ourselves to consideration of distribution of the single quantity. The question about the distribution of individuals over income x is one of most discussed subjects in econophysics (Yakovenko 2007).

At the end of the nineteenth century, sociologist and economist Vilfredo Pareto studied distribution of people of the various countries over income (Pareto 1897, pp 299–345) and found that function of density of distribution $p(x)$ over individual income x is defined by power function

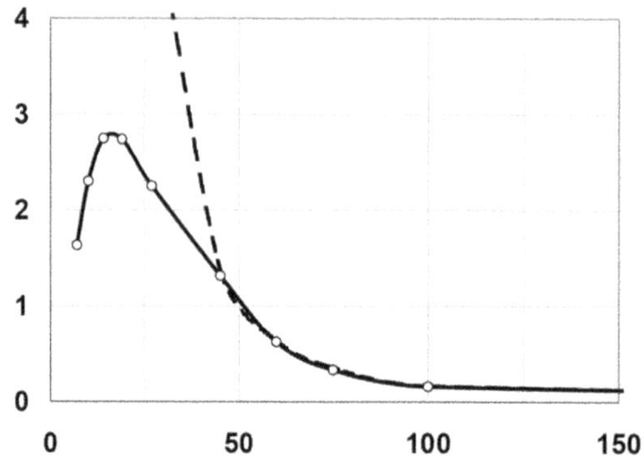

Figure 11.1. Density of distribution. The solid curve represents the probability of the density of the distribution for the income (presented in thousands of rubles) of the Russian population in 2018, calculated from available data (Rosstat 2019). The dotted curve represents the Pareto distribution (11.20) for $\alpha = 1.65$ and reproduces the real dependence for income $x > 45\,000$ rubles.

$$p(x) = Ax^{-(1+\alpha)}. \tag{11.20}$$

The quantity $p(x)\Delta x$ represents the number of individuals with an income between x and $x + \Delta x$. The index α in the power distribution (11.20) is called the Pareto index; the typical value of the index is $\alpha = 1.5$. Pareto believed that this value was universal, but further research did not confirm this (Souma 2002). The distribution (11.20), called the Pareto distribution, is not valid, of course, in the area of very small and large income values; there is a minimum income $x_{\min}$, below which survival of people is impossible, and some maximum income $x_{\max}$. Figure 11.1 illustrates the situation.

To norm the Pareto distribution (11.20) over the number of individuals, the areas with very small and very big amounts of the income can be excluded, and one gets the connection of the constant A with the total number of individuals N

$$N = A\int_{x_{\min}}^{x_{\max}} x^{-(1+\alpha)}\mathrm{d}x = \frac{A}{\alpha}(x_{\min}^{-\alpha} - x_{\max}^{-\alpha}). \tag{11.21}$$

It is not difficult to calculate the income received by all individuals

$$X = \int_{x_{\min}}^{x_{\max}} xp(x)\mathrm{d}x = \frac{A}{\alpha - 1}(x_{\min}^{1-\alpha} - x_{\max}^{1-\alpha}). \tag{11.22}$$

Expressions (11.21) and (11.22) allow calculating the average income on one individual X/N. Considering the quantity $x_{\min}/x_{\max}$ to be small, one can obtain in the simplest approximation

$$\bar{x} = \frac{\alpha}{\alpha - 1}x_{\min}. \tag{11.23}$$

At typical value $\alpha = 1.5$, the average income exceeds the living minimum $x_{\min}$ by about three times.

Number of individuals with the income below an average and the aggregate profit of this group can be estimated according to formulas (11.21) and (11.22), at integration from $x_{\min}$ up to average value $\bar{x}$, instead of $x_{\max}$. At calculation of the quantities, we consider the upper limit $\bar{x}$ in the integrals to be very great, so that the quantity $x_{\min}/\bar{x}$ can be neglected, and we have

$$\frac{N_{x<\bar{x}}}{N} = 1 - \left(\frac{\alpha}{\alpha - 1}\right)^{-\alpha} \approx 0.8,$$

$$\frac{X_{x<\bar{x}}}{X} = 1 - \left(\frac{\alpha}{\alpha - 1}\right)^{1-\alpha} \approx 0.4.$$

Hereinafter, the numerical assessments are executed for typical value of the index $\alpha = 1.5$. The result shows essential non-uniformity of distribution of the income: eighty percent of individuals do not receive even half of cumulative income.

It is possible to choose a point of division with a greater value: for example, to use average value of the income for a group of individuals with income above the average value

$$\bar{\bar{x}} = \left(\frac{\alpha}{\alpha - 1}\right)^2 x_{\min}. \tag{11.24}$$

Also, as shown above, we find number of individuals with values of income less than this size and the aggregate profit of this group

$$\frac{N_{x<\bar{\bar{x}}}}{N} = 1 - \left(\frac{\alpha}{\alpha - 1}\right)^{-2\alpha} \approx 0.96,$$

$$\frac{X_{x<\bar{\bar{x}}}}{X} = 1 - \left(\frac{\alpha}{\alpha - 1}\right)^{2-2\alpha} \approx 0.67.$$

This result shows, that almost all population receives only about two thirds of total income.

It is possible to continue and define the third point

$$\bar{\bar{\bar{x}}} = \left(\frac{\alpha}{\alpha - 1}\right)^3 x_{\min}. \tag{11.25}$$

Appropriating calculations show that less than one percent of individuals receive 20% of the income.

In application to actual situations there is an indispensability of specification of value of Pareto's index and the function of density of Pareto's distribution, but the basic conclusion of numerous researches remains constant: that income and abundance in modern societies are distributed non-uniformly (Piketty 2014).

Derivation of the distribution function

Since the time of Pareto (1897), many researchers tried to understand, whether the power distribution appears to be the law of Nature, or whether it can be reduced to more fundamental principles. We shall follow (a little bit differently) Richmond and Solomon (2001) who have shown that the normal Gaussian distribution of probability lies in the foundation of power law.

The distribution function (11.20) has been appearing as the result of a multitude of elementary actions among economic agents, and the phenomenon can be interpreted from this point of view (Scafetta *et al* 2002, Yakovenko 2007). We consider income of a person x as a label of a person, the probability of distribution $p(t, x)$ is considered as a characteristic of the system of individuals. The quantity

$$p(t, x)dx$$

represents the number of individuals having an income between x and $x + dx$ in the moment of time t.

There is a method that allows one to derive a kinetic equation for distribution function, considering the elementary acts of transition from any state to another one (see, for example, Lifshitz and Pitaevskii 1981). We begin with the change of the distribution function that can be written as

$$\frac{\partial p(t, x)}{\partial t} = \int_{-\infty}^{+\infty} [w(a, x + a)p(t, x + a) - w(a, x)p(t, x)]da. \tag{11.26}$$

Here, the probability $w(a, x)da$ of transition of the individual from state x to state $x - a$ is introduced. One can assume that the probability function has the standard Gaussian form

$$w(a, \langle a\rangle, \langle a^2\rangle) = \left(\frac{1}{2\pi\langle a^2\rangle}\right)^{\frac{1}{2}} \exp\left[-\frac{(a - \langle a\rangle)^2}{\langle a^2\rangle}\right]. \tag{11.27}$$

This function depends on income x through mean value $\langle a\rangle$ and dispersion $\langle a^2\rangle$. It is natural to assume that the dispersion $\langle a^2\rangle$ is an increasing function of income in the area of positive values of x (otherwise, the exchanges for the agents with high income can be ineffective), whereas $\langle a^2\rangle = 0$ for non-positive values. The mean value of the quantity $\langle a\rangle$ can be considered to be independent of the variable x.

One can assume that the change of income a is much less than its average value of income, that $a \ll x$, By virtue of this, the first part of integrand expression can be replaced by expansion

$$w(a, x + a)p(t, x + a) \approx w(a, x)p(t, x) + a\frac{\partial(wp)}{\partial x}\bigg|_{a=0} + a^2\frac{\partial^2(wp)}{\partial^2 x}\bigg|_{a=0}.$$

Then, the variables a and x are independent, and, by virtue of this, the kinetic equation for the distribution function can be recorded as

$$\frac{\partial p(t, x)}{\partial t} = \frac{\partial}{\partial x}\left[\left(\langle a\rangle + \frac{\partial\langle a^2\rangle}{\partial x}\right)p + \langle a^2\rangle\frac{\partial p}{\partial x}\right]. \tag{11.28}$$

On the given assumptions, equation (11.28) is valid in the area of positive values of x.

One can assume that the stationary solution of equation (11.28) can have the form of function (11.20), which gives the relation between the parameters

$$1 + \alpha = \frac{x}{\langle a^2 \rangle}\left(\langle a \rangle + \frac{\partial \langle a^2 \rangle}{\partial x} \right). \tag{11.29}$$

One can note that, if the mean value $\langle a \rangle$ is constant and the dispersion $\langle a^2 \rangle$ is proportional to x, that is $\langle a^2 \rangle = kx$, the right-hand side of the relation (11.29) appears to be independent of x. Taking these suppositions into account, one can define the index of Pareto as

$$1 + \alpha = 1 + \frac{\langle a \rangle}{k}. \tag{11.30}$$

Our speculations reduce the power distribution to the underlying Gauss distribution, whereas the latter has to have two features that lead to the observed inequality. The first one is that the dispersion of the probability increases in proportion to income x, which is a condition of the existence of the system. The second feature is that the distribution function $w(x, a)$ is non-symmetric in relation to the variable a. In that case, if the function $w(x, a)$ is symmetric with respect to the variable a, $\langle a \rangle = 0$ and the dispersion of the probability does not depend on x, the first term in the right-hand side of equation (11.28) disappears, and the kinetic equation reduces to the one-dimensional diffusion equation.

References

Brandmeir K, Grimm M, Heise M and Holzhausen A 2011 *Allianz Global Wealth Report 2011* (Munich: Allianz SE, Economic Research & Corporate Development)

Bukharin S N and Yu M S 2014 *Evolyutsiya elity -materialy i issledovaniya [Evolution of Elite–Materials and Researches]* (Moscow: Akademicheskiy proekt, Gaudeamus)

Dugin A G 2011 *Etnosociologija [The Etnosociology]* (Moscow: Akademicheskij Proekt)

Ellerman D 2007 On the role of capital in 'Capitalist' and in labor-managed firms *Rev. Radic. Polit. Econ.* **39** 5–26

Ellerman D 2015 On the renting of persons: The neo-abolitionist case against today's peculiar institution *Econ. Thought* **4** 1–20 http://www.worldeconomicsassociation.org/files/journals/economicthought/WEA-ET-4-1-Ellerman.pdf

Feinberg E L 1987 *Art in the Science Dominated World: Science, Logic, and Art, Studies in Cybernetics* vol 13 (New York: Gordon and Breach)

Feinberg E L 2004 *Dve kul'tury: Intuitsiya i logika v iskusstve i nauke [Two Cultures: Intuition and Logic in the Art and Science]* 3rd edn revised and extended (Fryazino: Vek 2)

Gorskii D P 1985 *Obobshchenie i poznanie [Generalization and Understanding]* (Moscow: Mysl')

Grinberg R and Rubinstein A 2005 *Economic Sociodynamics* (Berlin: Springer)

Gumilev L N 1990 *Ethnogenesis and the Biosphere* (Moscow: Progress)

Kalyuzhny O N 2005 *O rekursivnosti psikhicheskogo protsessa prinyatiya resheniya: Zadacha o 'khoroshey' i 'plokhoy' kuchkakh [About Recursiveness of Mental Process of Decision-Making: The Problem about 'Good' and 'Bad' Batches]* http://www.oleg.314159.ru/

Keller M S 2007 Why is music so ideological, why do totalitarian states take it so seriously: a personal view from history, and the social sciences *J. Musicol. Res.* **26** 91–122

Kordonskiy S G 2006 *Rynki vlasti: Administrativnye rynki SSSR i Rossii [The Markets of Authority: Administrative Markets of the USSR and Russia]* (Moscow: OGI)

Kordonskiy S G 2008 *Soslovnaya struktura postsovetskoy Rossii [The Class Structure of the Post-Soviet Russia]* (Moscow: Institut Fonda 'Obshchestvennoe mnenie')

Lefebvre V A 1990 *The Fundamental Structures of Human Reflection* (New York: Peter Lang Publishing)

Leontief W W 1986 *Input-Output Economics* 2nd edn (New York: Oxford University Press)

Lifshitz E M and Pitaevskii L P 1981 *Physical Kinetics* (Oxford: Pergamon)

Mednikov B M 1976 Genom i jazyk: paralleli mezhdu jevoljucionnoj genetikoj i sravnitel'nym jazykoznaniem [Gene and Language: Parallels between Evolutionary Genetics and Comparative Linguistics] *Bjulleten MOIP. Otd. biol. [Bull. Moscow Soc. Expl. Nat., Biol. Issue]* **81** 134–47

Menger K 1938 An Exact Theory of Social Groups and Relations *Am. J. Sociol.* **43** 790–8

Menger K 1983 On social groups and relations *Math. Soc. Sci.* **6** 13–23

Pareto V 1897 *Cours d'economique politique* 1st edn (London: Macmillan) Also:
Pareto V 1964 *Cours d'Economie Politique* Nouvelle edition par G H Bousquet and G Busino (Geneva: Librairie Droz)

Piketty T 2014 *Capital in the Twenty-First Century* (Cambridge, MA: Harvard University Press)

Plushchevsky A M 2001–7 *Termodinamika istoricheskoy zhizni narodov: Matematicheskoe modelirovanie istoricheskikh protsessov s ispol'zovaniem passionarnoy teorii etnogeneza i statisticheskoy termodinamiki [Thermodynamics of the Historical Life of People: Mathematical Modelling of Historical Processes with the Use of the Passionary Theories of Ethnogenesis and Statistical Thermodynamics]* http://ampsoftware.narod.ru/model.htm

Pokrovskii V N 2018 *Econodynamics: The Theory of Social Production* 3rd edn (Dordrecht: Springer)

Richmond P and Solomon S 2001 Power laws are disguised Boltzmann laws *Int. J. Mod. Phys.* C **12** 333–43

Rosstat 2019 *Distribution of the Population by the Amount of Per Capita Monetary Income* Table: urov_31g.doc (updated on 26.12.2019) Available at: https://http://www.gks.ru/ (accessed February 2020)

Samuelson P A 1954 The pure theory of public expenditure *Rev. Econ. Stat.* **36** 388

Scafetta N, Picozzi S and West B J 2002 *Pareto's Law: A Model of Human Sharing and Creativity* (arXiv: cond-mat/0209373v1)

Schumpeter J A 1911 *Theorie der wirtschaftlichen Entwicklung: Eine Untersuchung über Unternehmergewinn, Kapital, Kredit, Zins und den Kojunkturzyklus* (Berlin: Dunker und Humblot)

Shorrocks A, Davies J B and Lluberas R 2010 *Global Wealth Databook 2010* (Zurich: Credit Suise Group AG)

Smith A 1976 *An Inquiry into the Nature and Causes of the Wealth of Nations* (Oxford: Clarendon)

Souma W 2002 *Physics of Personal Income* (arXiv: cond-mat/0202388v1)

Turner J H 1974 *The Structure of Sociological Theory* (Homewood, IL: Dorse)

Walras L 1874 *Elements daeconomie politique pure ou theorie de la richesse sociale* (Lausanne: Corbaz)

Weidlich W 1994 Synergetic modelling concepts for sociodynamics with application to collective political opinion formation *J. Math. Sociol.* **18** 267–91

Weidlich W 2000 *Sociodynamics: A Systematic Approach to Mathematical Modelling in the Social Sciences* (Amsterdam: Harwood Academic)

Yakovenko V M 2007 *Econophysics, statistical mechanics approach to in Encyclopedia of Complexity and System Science* (Berlin, Heidelberg: Springer)

Zinoviev A A 2002 *Logicheskaya sotsiologiya [The Logical Sociology]* (Moscow: Socium)

Chapter 12

Dynamics of cognition

Thermodynamics, as well as any scientific discipline, appears to us as an artificial system of concepts, relations and principles that exists in our imagination. We believe that this is the reality we are dealing with. However, as philosophers teach us, this is not the reality, but only our image of reality. In our knowledge, we assign to objects the properties of substantiality, causality, etc—all these properties are only categories of our mind. What the world is like independently of us, we don't know; but the world we know is our own creation, the product of the cognizing subject, and it is interesting to understand what urges us to design such reflections.

12.1 Imperative to putting in order

It is believed that in the process of evolution, a human body has acquired the ability to perceive signals from the environment and organize them, which allows a person to have a generalized idea of the environment and make decisions about the best behavior. According to the words of one of the creators of engineering psychology (Craik 1943, chapter 5): 'If the organism carries a 'small-scale model' of external reality and of its own possible actions within its head, it is able to try out various alternatives, conclude which is the best of them, react to future situations before they arise, utilize the knowledge of past events in dealing with the present and future, and in every way to react in a much fuller, safer, and more competent manner to the emergencies which face it'. It is remarkable that the ability to order external signals has developed as one of the mechanisms of adaptation of organisms for life in the environment and remains embedded in the nature of humans and animals as an instinct for ordering. The processes of ordering impressions on the unconscious and conscious levels determine the chain: signals–images–concepts–statements–events–theories.

doi:10.1088/978-0-7503-3451-8ch12

12.1.1 Level of unconscious mind

We should think that the chaos of signals is perceived by both a human and any animal, but without pre-processing and creating generalized *images*, it would be difficult to use these signals. This work is performed by the nervous system and brain of the biological organism (Haken and Haken-Krell 1992, Turchin 2000). It should be assumed that both animals and humans have an innate signal processing mechanism at the unconscious level—some information device similar to a computer (Lilly 1987, Hopfield 1999), with a built-in program for transitions from one state to another. When signals are received, the system is excited, and a process begins that moves the system from the excited state to the final state, which is assumed to be stable in at least some respects. This process allows the organism to develop generalized characteristics (images) of a set of impressions (signals).

There is also some internal program—an innate mechanism for recognizing similarities and differences of the *images*, which can be qualified as an ordering instinct. With direct perception of images A, B, C, ⋯ a common feature is captured, which gets a name—a label, the term α, which fixes the representation of the *concept. The designation of the concept* can be considered as an elementary element of description of the external world. Any designation is an abstract construction, and there is nothing in reality that corresponds to the concept as such, but, however, the concept reflects the idea of some common features of the few (or many) *denotata*.

Both humans and animals acquire some experience during their lives based on the existence of an unconscious process of ordering signals. The work of the nervous system and brain takes place on an instinctive, unconscious level. The owner of the biological 'computer' has no control over the dynamics that are determined by the rules of transition from one state to another. Fundamental concepts and certain relationships between concepts appear at the level of the unconscious.

12.1.2 Level of consciousness

The main work on the ordering of influences on the body and the formation of concepts proceeds without awareness of what is happening, as they say, at the level of the unconscious. In this case, *the concepts* arise, which become facts of our consciousness, they can be thought about and combined. Each concept is understood as a unity of label and meaning. The establishment of new relations between concepts is recorded as the acquisition and expansion of meaning. Progress in the assimilation of the Universe is accompanied by the differentiation of concepts, the introduction of new concepts; there are concepts that express the relationship between concepts. People's experiences are registered as systems of concepts, as systems of statements.

At the level of consciousness, the rules of ordering are formulated, like the rules of operating with labels and reasons. The laws of formal logic, described for the first time by Aristotle, are the rules for ordering our conscious impressions, which, however, cannot lead beyond the existing system of concepts and statements[1]. New concepts and statements arise at the informal level.

Fundamentally new concepts and statements are formulated at the level of the unconsciousness with the participation of our conscious mind. There are practically no purely logical discoveries (Hadamard 1951). All discoveries are the result of the insight, which is the result of the collaboration of our conscious and unconscious minds. Rules were developed to facilitate and stimulate the release of our conscious mind from the fetters of logical reasoning and to appeal to the level of the unconscious (Altschuler 1979, De Bono 1970).

The conscious and unconscious interact and support each other. There is no sharp and precise boundary between the conscious and the unconscious, and the successive layers of the unconscious can be distinguished. According to Sigmund Freud, there is a level of pre-conscious—a certain border area that, under certain conditions, can be opened to conscious contemplation. A Russian scientist Nalimov (1989, p 104) highlights the level of pre-thinking, and offers an expanded map of consciousness that allows him to approach the explanation of the rules of ordering (probabilistic rules of finding meaning) on the unconscious level.

12.1.3 Statements we believe in

As a result of interaction of the conscious and unconscious levels of our mind, statements are formulated, some of which are recognized as true ones. Such statements are considered to be *facts*. A person recognizes the facts directly (as a result of the unconscious work of the nervous system and brain) or after reasoning with the help of formal logic. Knowledge of the facts is vital for a person, when he acts and plans his activities. All facts constitute a reality that is obviously different for different people. The reality of a person appears and disappears along with the personality.

The carriers of consciousness are individuals with limited abilities, but the existence of labels of concepts (language) makes it possible to make concepts and statements to be facts of public consciousness. There is an intense exchange of messages among the individuals of the society that leads to the emergence of something whole—a collective, social consciousness that includes all the facts available to any person. The totality of all facts in the public consciousness is a set that includes the facts of any individual, so that the reality perceived by the society, is greater than the reality of any human, but any fact in the public consciousness, in principle, is available to any individual. We can assume that the hypothetical actual reality is more than the reality we are dealing with. But how big is it and is there anything outside of the reality we are dealing with?

The totality of facts is a record of people's collective experience. It is impossible to list all the facts that a person faces in his life. Generalization of human experience is unthinkable outside of the already existing system of knowledge, which is an

[1] An introducing presentation of logic in modern terms can be found, for example, in the book of Segeth (1973).

accumulation of previous experience and is relatively *a priori* for the researcher. There is a continuous process of cognition—the inclusion of new facts in the system of knowledge. At the level of personality, a person subconsciously tries to find the proper place for each new fact in the system of facts (to understand a new fact), and a person feels satisfied when all the facts are compatible with each other in the system. Otherwise, there is a feeling of discomfort and excitement. This is a manifestation of the imperative of putting in order.

12.1.4 The events

Everything that happens in the Universe, including us directly, is interconnected, but, in our minds, we separate what we think about reality from the rest of the Universe. From time to time, our attention is focused on certain phenomena that attract our attention. We may even believe that these phenomena exist separately from the rest of the Universe, which, however, has a direct influence on the phenomena in question. The phenomenon thus highlighted can be called an *event*, and everything except this event is treated as the *environment*. The events are described using a system of concepts, but we are not given any external language to describe an event: a concept language can only be developed, when the event is being considered, and we have to deal with a set of events, even if only one event needs to be described.

A certain event and an uncertain environment are the initial concepts of our approach to description or modeling: these concepts are often replaced by others: what we pay attention to can be called: an object, a process, a system, and so on. For certainty, it is convenient to choose one word; let's agree to call it *event*—this is what we are trying to describe or, in other words, model.

How can I explain what is an event? An event is something that actually attracts our attention. The event stands out in reality in some way, which is why the event attracts our attention. However, in reality there are no events in the forms we imagine them to be; it is our attention and imagination that create the event. The event is something that is described and its description simultaneously, although there is some opposition and difference between the event and its description. The only essential property of an event is its simultaneous existence in two worlds: the material and the ideal. However, there is only one world—the Universe and the division into two worlds is necessary only to describe the process of understanding.

We can imagine the Universe as a collection of events. The real Universe is different from our imagination: it is not made up of events and obviously always turns out to be more than we can imagine. However, the Universe as a collection of events seems to us a good approximation. Each event can be defined by its relationships with other events. Thus, a set of relationships defines an event. We don't know what events are, and we don't know what the Universe is. However, the statement 'the Universe is a collection of events' seems to us to have some significance. The meaning of this statement is that the Universe is structured, has a certain architecture, and you can choose a certain event in the Universe. In any case, we feel that there is some distinction and opposition between the model and the modeling object. At the same time, it is something that must be considered together.

12.2 Description of the events

An event that attracts our attention and is described using intuitive concepts and statements is a subject of interest to researchers. Could we talk about any general features of events already at this stage of the transition from intuitive selection to conscious description?

12.2.1 Space and time

Researchers record the multiplicity of events in the Universe; at every moment between the past and the future, we record a set of objects or bodies in our Universe, these objects can be spatially ordered using three parameters, and they say that all bodies are placed in a space of the three dimensions. The body we are considering may represent a complex spatial formation, and it is naturally distinguished by the fact that the connections within the body are more significant than the connections of the selected body with all other bodies. To characterize the state of a body, it is assigned the parameters w_j ($j = 1, 2, \ldots r$). Each body does not exist without its inherent motion; to describe changes in its state, we need another universal parameter—time t. So that the event that we are exploring is defined by functions of time

$$w_j = w_j(t, w_1, w_2, \ldots w_{j-1}, w_{j+1}, \ldots w_{r-1}, w_r), \quad j = 1, 2, \ldots r. \tag{12.1}$$

These functions include, of course, spatial coordinates (for describing the architecture) and parameters that define interaction with the environment.

So, the descriptions of events are based on *a priori* given classical space–time. In the simplest case, when the event body is a material point of mechanics, the position of the event is fixed by setting three spatial coordinates x, y, z and time t; the state of the body in question is determined by velocity as a function of the listed variables. It is believed that all events, including the most improbable, are developing in the space of four dimensions (Krasnikov 2015). Three spatial coordinates and time are necessary to indicate the location of any event among other events, so we can say that the observed Universe is an ordered set of events. In another way, we can say that the relationship between the elements of the Universe is such that the four parameters listed above are necessary to describe the interactions. Just as in the simplest case, when the *greater–less* relation arranges all numbers in a row, induces a one-parameter order, and leads to a geometry with a space of dimension one, the *earlier (before)–later (after)* relation orders a set of events and defines the geometry—the four-parameter geometry of space–time, as shown by researchers (Robb 1936, Pimenov 1991).

The idea of *a priori* four-dimensional space–time, against which all events are developing, does not face any misunderstandings when describing large-scale events, but, when turning to the dynamics of the simplest elementary particles, one needs to assume existence of, in addition to four, additional dimensions, which are manifested at such small distances that they cannot be detected properly[2]. It seems that in the microcosm we encounter the most profound properties of the Universe, and it is at this level that we can understand why we need exactly four parameters to establish order in our observations, which we are now forced to accept as an empirical fact.

The fundamental theory of space–time is based on the concept of long-range action (Kulakov 2004, Vladimirov 2009, 2015). The consideration begins with postulating a set of elements of indeterminate nature, and postulating numerical relations between the elements of this set (unary relations), which determine the laws of interaction of elements. One can imagine that as a result of some processes, a given set of elements turns into another set with the same or a different number of elements. Considering binary relations between elements of two sets allows formulating the laws of 'movement' of elements and defining the geometry of 'space–time' (Vladimirov 2009, 2015). Two sets of three elements forming a binary structure of rank (3.3) define the geometry of a classical four-dimensional space–time. Consideration of the rank structure (4.4) leads to a geometry of the dimension nine with a cubic metric definition. We can assume that our four-dimensional Universe is somehow embedded in the space of nine and more dimensions, and events in our Universe could be considered only as projections of events in a larger world. With the five or more additional coordinates, we can observe various miracles in our Universe, the manifestation of which we meet, first of all, in the microcosm. But it cannot be excluded the possibility that in our macrocosm may occur 'miraculous phenomena', reports of which sometimes appear in print (Dunne, 1927, Bekhtereva 1999, chapter 'Through the Looking Glass'). The existence of a transcendental space seems to explain miraculous phenomena.

In the meantime, we are returning to a single Universe with a set of events, where relations are set that induce the known geometry of space–time, and we need four parameters to indicate the place of the event among other events. Surprisingly, in our world, the temporal order is independent of the spatial order with very high accuracy, so we can assume two types of relationships. The relations of sequence induce an order that we call time, the relations of interaction define a three-parameter space.

12.2.2 Principle of self-preservation

The phenomenological event is set by the state parameters $w_j (j = 1, 2, \ldots r)$, which change due to internal and external circumstances, so that the event can be associated with a set of functions (12.1). These functions, of course, may not be explicitly defined, but it is assumed that there are some dependences.

It can be assumed, based on observations and experience, that if nothing acts on the event body, the event body retains its state. This state can be considered as a natural state of the event body, in which all state parameters are constant. For example, the natural state of the event body: for a classical particle, the velocity is

[2] An overview and evaluation of various assumptions can be found in the book of Vladimirov (2015).

constant v = const.. This is movement by inertia, when you cannot tell whether it is movement or rest. For a thermodynamic system in equilibrium, the state functions are constant E = const., S = const..

Obviously, the natural state is some idealization of the event; we cannot get rid of external influences. There is an interaction of bodies, but if we want to consider a single event, we must introduce a measure of the influence of external events on the event in question. The transition from one state to another is carried out under the influence of an external influence—or of some appropriately defined 'force'. However, a certain event can only affect events that occur later than the first one, but not every preceding event really affects the event in question.

Curiously, it is possible to indicate the general nature of the reaction of the event body to external influences: the body tends to reduce the impact result. As an example, you can specify, first of all, the movement of the body: when an external influence appears, the force of inertia emerges. Another example: a thermodynamic system in equilibrium that is affected responds (changes) in such a way that the effect of influence is reduced to a minimum—the Le Chatelier principle. It is known that biological objects react to external influences in such a way as to ensure survival: the preservation of the organism and the species (see section 9.3.1).

Thus, when considering an event, one should always take into account an important principle, which can be called the principle of causation, or the principle of inertia, or the principle of self-preservation.

12.2.3 The picture of our world

As an event, we consider the movement of an elementary particle, a change in the thermodynamic system, and the development of a human population. Processes involving elementary particles are observed at microscopic level with scales of the order of 10^{-15} cm and times of the order of 10^{-22} s. On the other hand, cosmology leads us to times of the order of 10^{10} years (the 'age' of the Universe) and to distances of the order of 10^{28} cm (the distance to the event horizon, that is, the farthest distance from which a physical signal can be detected). We do not know whether our Universe is infinite or not, and we cannot rule out the assumption that events in our Universe are only a four-dimensional projection of events occurring in the space of a larger number of dimensions. In our Universe, in our epoch, four architectural levels of the events under consideration are naturally distinguished:

- elementary particles;
- atoms and molecules;
- simple and complex thermodynamic systems;
- supra-organismal systems.

Events are developing in space and time; the characteristics of some events are shown in table 12.1.

The body of each subsequent event in the list can be reduced to a set of event bodies from previous levels, but this does not mean that the behavior of a more complex event can be reduced to the laws of behavior of 'elementary' events. There

Table 12.1. Characteristic parameters of some events.

Body of the event	Linear size (m)	Time of existence (year)	Constituents
Nucleus of the atom	Radius 10^{-15}		
Atom	Radius 10^{-12}		
Virus	Length 10^{-7}		10^{8}–10^{10} atoms
Light	Wavelength 5×10^{-7}		
Human	Height 1.5	10^{2}, as an individual	10^{16} cells
The Earth	Radius 6×10^{6}	10^{6}, as a species 5×10^{9}	10^{28}–10^{30} atoms 4×10^{51} protons and neutrons
The Earth system	To the Moon 4×10^{8}		
Solar system	Radius 6×10^{12}		
Galaxy	Radius 10^{20}		
The Universe	Radius 10^{26}	2×10^{10}	10^{80} protons and neutrons

is no fundamental, elementary level of event description, although, as Niels Bohr states, there are elementary concepts and fundamental principles. Although the thermodynamic system, for example, consists of atoms and molecules, the behavior of the macrosystem is not completely reduced to the behavior of atoms and molecules, the thermodynamic principles of the system's behavior are formulated independently.

12.3 Mathematical modeling

The observed phenomena were described in previous chapters using universal mathematical concepts and constructs, which can be defined as *mathematical modeling* of reality, and now we will turn to the description of this method, which is used in various disciplines. First of all, what is a *mathematical model*?

Of course, you can find many definitions of the concept. For example, in the book of Karlin (1959), you can read on page 1: 'A model is a suitable abstraction of reality preserving the essential structure of the problem in such a way that its analysis affords insight into both the original concrete situation and other situations which have the same formal structure'.

This could be considered an excellent definition, if we first knew what *is a real phenomenon* (in similar definitions, one also uses other terms: system, process, phenomenon, situation, and many similar ones). Further, how can we judge that the model 'preserves the essential structure of the problem'? When we think about it, we realize that the real phenomenon denotes our idea of a 'real phenomenon', described in terms that we understand, and our desire to write down a mathematical model can only be interpreted as a desire to clarify some details in the existing description of a 'real phenomenon'.

Mathematical modeling turns out to be one of the methods of constructing our representations of the reality. The ultimate goal of mathematical modeling is to establish principles that allow us to organize the known facts and predict new facts that could be included in the knowledge system. Mathematical modeling is a universal way of describing reality, which gives us the illusion of understanding. In fact, we can only describe reality bringing one event to another, linking together a variety of events ⋯. Mathematical models are a description of the relationship between events, and researchers are forced to abandon the illusory aim of 'understanding phenomena'. Performing the task of mathematical modeling is the work of an applied mathematician, who, in contrast to the 'pure' mathematician, according to Birkhoff (1969), 'cares about the meaning of his words and symbols'. Applied mathematics deals with mathematical constructions as models of 'real objects'.

12.3.1 Description tools

Mathematical modeling deals with the description of events using mathematical concepts, and, as a rule, the researcher finds all the mathematical objects necessary for building a model in a ready-made form. The tools for describing events are provided by mathematics, which studies special logical structures—mathematical structures (elements and relations) (Yaglom 1980). This is the subject of pure mathematics (as opposed to applied mathematics).

Pure mathematics studies the relationships between various mathematical elements. For example, there is a certain relationship between two mathematical objects: a function

$$y = \sin t$$

and the differential equation

$$\frac{d^2y}{dt} + y = 0.$$

Indeed, the function is a solution of the differential equation corresponding to the initial conditions

$$y(0) = 0, \quad y'(0) = 0.$$

A mathematical model is a mathematical structure that, regardless of the meaningful contents of the model, must meet formal requirements such as the conditions for the existence and stability of the solution, self-consistency, and consistency with other models.

12.3.2 Scheme of the process of cognition

Mathematical modeling starts when a proper mathematical object is compared to some element of reality—a number, first of all, then the relationship between the numbers—functions, then the relationship between a certain quantity and the rate of

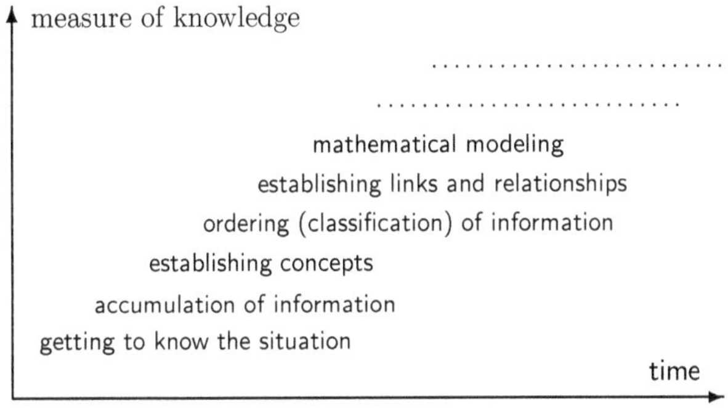

Figure 12.1. The hard way of cognition. The cognition begins with simple description and, passing through the formalization stage, never comes to its end.

change of this quantity—a differential equation, and so on. This is amazing, that all mathematical objects, even quite complex structures such as matrices, spinors, and complex numbers, are used to construct mathematical models. A constructed mathematical object is called a mathematical model, which defines a coherent system of concepts and creates a language that allows us to describe events and discuss related problems. A mathematical model is an artificial object in our world (Simon 1962).

Let us try to describe the sequential steps of a researcher who, for the first time, encounters a previously unknown phenomenon and seeks to construct a mathematical model. These steps are also shown in figure 12.1.

1. A situation that should be understood occurs when you feel that there is a problem that you would like to solve. For example, there is a change in a certain quantity (the position of the body or the output of the production system) over time, and the researcher tries to understand what the changes are related to. There may be a contradiction in some facts. Usually the problem is formulated in the most general way: 'Act in such a way, please, to make us happy'.

2. Selecting an event from the environment is the result of a preliminary study of the situation. It turns out what is essential and what can be omitted in the situation under consideration. Concepts and essential connections with the environment are established. As Thomas Carlyle (1837, chapter 3.5.I) writes about social phenomena: 'Any approximation to the right Name has value: were the right name itself once here, the Thing is known thenceforth; the Thing is then ours, and can be dealt with.' A descriptive check-in begins.

3. Schematization of an event means the introduction of a consistent system of concepts and symbols to denote significant quantities and the assumed relationships between quantities.

4. The purpose and method of modeling. The researcher thinks about the problem that needs to be solved and the results that can be obtained. Each specific situation requires a specific approach. There are no methods that are suitable for all cases.

Methods are developed by those who study special situations: methods appear for a very specific reason. It should be noted that a mathematical model is an immaterial artificial object, and it is possible to consider the construction of the model as the design and construction of an artificial object.

5. Mathematical model: first iteration. Finally, some relations between quantities are formulated in symbolic form. In the simplest case, the change in body position can be measured at consecutive points of times. Then you can find a function whose values match the fixed values. A function is, of course, a mathematical model of a phenomenon that can thus be interpolated between fixed moments of time and extrapolated outside of the time interval in question. To get a more reliable extrapolation, you can use the growth rate of the value. The mathematical model takes the form of a differential equation. It may happen that you need to take into account the growth rate of the growth rate to find a more appropriate model, and so on. The latter cases assume that the mathematical model contains higher-order derivatives. This is equivalent to a system of differential equations for multiple variables. How many variables are needed to describe the phenomenon properly? From the possible non-equivalent methods of description, the most adequate model is chosen, which receives the status of a theory of the phenomenon. The first iteration is usually the most difficult step from recognizing the existence of an event to a mathematical model.

6. Model analysis clarifies the meaning of values and corrects the relationships between values; it stresses the main properties of the model. The analysis checks whether the model meets the fundamental principles (conditions for the existence and stability of the solution, self-consistency, consistency, and so on).

7. Comparison with empirical data finds out the advantages and disadvantages of the model. A mathematical model is a mathematical object, and the question arises about its relationship to the modeled object. Should we demand that the model correspond to reality? Perhaps, if we knew what reality is, but we have to say no, because we don't know what reality is. We have no reality other than a system of concepts and mathematical models. It is impossible to separate the doctrine of reality and the doctrine of mathematical modeling of reality. Can the reader imagine a situation where physics students are taught first 'physics' and then 'mathematical models in physics'? However, this situation in another field does not look wild: some students are taught 'economics', and others 'mathematical models in economics'. This is evidence of a lower level of development of the system of economic concepts and models in comparison with the system of concepts and models in physics.

8. The conditions of applicability of the model. None of the models is universal, so that to avoid confusion, it is important to specify the areas of applicability of the model.

9. The consistency with other models. Should we require (in the case, when there is a developed system of mathematical models) that the model corresponds to all known mathematical models? This is impossible, because we know that some well-known models turn out to be inadequate, that is, they definitely do not include known facts in the description.

10. Mathematical model: second iteration. Unlike the first iteration, the second and subsequent iterations have a predecessor—a mathematical model. Under each iteration, the model is improved, analyzed, and compared with new facts. The number of iterations is unlimited: any model is approximate and can be refined. It is better to start with a simple model than to try hard to build a good model in all respects. That is why I have given a scheme of modeling that is not good in all respects, but it can be used as a starting point.

Simulation

In some situations, a mathematical model cannot be realized and studied due to some difficulties. Simulation is used when the details of an event are not sufficiently formalized, such as when a Brownian particle randomly wanders with no intersection, or when there are no other means to solve the problem, such as when studying flows in continuous media, or when studying the dynamics of radiation destruction of a crystal lattice. In this case, a simulation model can be built: a modeling algorithm is being written, which implements a computer simulation of elementary processes that make up the essence of the system under study, while preserving their logical structure, the sequence of time and the composition of data of the state of the system. In simulation, we must imagine a mathematical model, at least in its most general form. Simulation modeling concretizes and complements the object model.

12.3.3 The system of physics models

One can consider the mathematical model as a condensed form of our knowledge, it is a certificate in what we have studied, know and, as a result, can control any phenomenon. At the same time, the mathematical model is a measure of our ignorance. The social function of a mathematical model is to store our knowledge in a compact form.

It should be noted that the reflection of reality gives not one model, but a system of mutually consistent models, and to conclude this section we mention an example of a system of models that can serve as a sample, the system of models of physics, which was the most developed, so that physics, more exactly, theoretical physics could be called a school of mathematical models of natural phenomena. First used to describe the phenomena of inanimate nature, the methods of theoretical physics are successfully applied to describe the phenomena of living nature and social phenomena. Theoretical physics has become synonymous with applied mathematics.

The unique ten-volume course in theoretical physics by Landau and Lifshitz, published and reprinted in Russia since 1936, contains a huge amount of contents due to the lapidary, but very informative style of presentation, so that these volumes are considered an encyclopedia of theoretical physics. The description of a phenomenon always begins with postulating, perhaps sometimes obvious statements, thereby achieving an amazing transparency of the description; the presentation involves the reader in the study of the phenomenon, being a guide to action. Often, known results are derived using simpler and clearer methods. The Course had a strong influence on many researchers, including the writer of these lines, and in his books the author unwittingly reproduced the style of the Course.

In previous years, students in the Soviet Union were told: look at the Course, you can find everything there, but now, probably, you have to turn to other books to find the latest developments: in recent years explored new phenomena, sometimes going beyond the 'traditional physics', the researchers seek to find the principles on the basis, which allows presenting fundamental topics of theoretical physics to build in a more compact form (Vladimirov 2009, Bychkov and Zaytsev 2019). It is now becoming clear that 'big physics' includes the study of not only biological, but also social problems.

Based on the empirics, physical theories are being formulated as axiomatic theories: they are based on a set of statements, axioms often not obvious, but, nevertheless, in their consequences compatible with the facts. Examples are mechanics, the theory of electromagnetism, and thermodynamics. A fundamental feature of physical models is that all events are assumed to unfold in time and space, which are, nevertheless, themselves objects of research. Features of space–time are related to the nature of physical laws (Mach 1883, 1906, Kulakov 2004, Vladimirov 2009, 2015).

Mathematical models of physics are based on heuristic visualization, we can say heuristic principles and heuristic models. By the latter we mean the objects and structures that physicists study, such as an ideal gas, a perfect crystal, and many others. Heuristic principles include:

Principle of causality and self-preservation. This principle has already been discussed in section 12.2.2.

Principle of symmetry (covariance). Equations should not change their meaning during similarity transformations. An example of application of the principle is presented in section 7.4.

Principle of correspondence. If the event can be viewed from different points of view, for example, as a whole (a phenomenological model) or with the consideration of a detailed structure (a heuristic model), there must be a correspondence in the results.

Principle of similarity and dimension compliance is discussed in detail, for example, in the book by G I Barenblatt (1982).

Principle of beauty and harmony. Surprisingly, adequate theories turn out to be elegant, and their contemplation evokes feelings similar to those that arise when contemplating a good picture (Chandrasekhar 1987). The right theory must be elegant.

The knowledge of the principles and axiomatic nature of physical theories allows us to get some statements deductively, using clear logical rules. However, physical theories are not strictly deductive; physicists often use inconsistent theories.

12.4 Science as a social institution

We think about science as about a system of concepts or as about a system of models reflecting the reality to which we belong. The scientific activities include the analysis of facts, the invention of models, and the establishment of relations between models $\cdots$. In this respect, science is a developing self-organizing system of knowledge. There are no *a priori* principles, the principles are formulated during the development. However, science is an important social institution, and, as a social institution, science is supported by society and has, we should think, a special social function. What is it exactly?

12.4.1 Social function of science

It is sometimes said that science explains the Universe and provides insight. But a careful examination of scientific achievements shows that science has no real basis that allows science to explain anything. Science, or rather people who serve in science, explores fragments of reality, tries to connect different fragments with each other and reconstruct the whole. The process is not complete and it is not clear whether it can be completed. Of course, we can say that understanding is achieved when one fragment of reality is linked to another fragment, but in the long run this is just a manifestation of the imperative of ordering.

Science collects and describes facts, organizes the collected facts. It is known that samples of successful classification of facts are evaluated as significant achievements of science. As examples, we can recall the classification of plants in *Systema Naturae* (1735), *Fundamenta Botanica* (1736) and *Species Plantarum* (1753), performed by the Swedish botanist and researcher Carl Linnaeus (Carolus Linnaeus), and the principle of periodicity (the periodic table) of elements (1869) established by the Russian chemist Dmitry Mendeleev.

The ordering of facts sometimes is performed by very clever and complex methods: some abstract concepts have been invented for this purpose, for example, energy, entropy, value. To find additional facts that researchers need to fill in the gaps in our knowledge, special complex and sometimes very expensive enterprises are organized: think, for example, about modern colliders. But the ultimate goal of the search is not facts, but principles of ordering.

Science is a social institution that has the function of ordering of the people's experience. Science collects and orders facts, classifies events, introduces new concepts and constructs theories, with logic and intuition working together. After all, science develops, approves, and transmits the principles of ordering that are spoken of as laws of nature. But these are not laws of nature, but only heuristic principles of ordering.

As an irreplaceable social institution, science has been and is a part of the social organization at all times of human existence: from the stone age to the present day. Understood in a broad sense, science begins simultaneously with the beginning of the society, and the development of science is based on the imperative of order. Even in ancient times, the ordering of random observations led to a system of thoughts about nature and people—magic, which can be considered, if not as the first scientific system, then the direct predecessor of science as it is stated by Frazer (1922). The method of constructing scientific systems was invented by the ancient Greeks (see, for example, chapter 2 in the book of Yaglom 1980). It is believed that modern systems of natural science date back to the 16th century. In our time, science is an element of the social system, along with other institutions: the army, the police, the education system ⋯.

12.4.2 About scientific method

Science begins with a noble concern about the inconsistency of facts and phenomena. Our minds prefer to deal with theories and general laws rather than large accumulations of unrelated data that are difficult to remember. Different theories organize the world in different ways according to the type of problems they propose to solve. The highest value of scientific theories is 'their correspondence with reality'. According to Popper (1959, 1963), a theory can be considered scientific, if the methods for testing it are specified.

In its simplest form, the emergence of a scientific theory can be described as follows. To link (explain) certain phenomena, some assumption is made—a hypothesis appears. The consequences of this hypothesis are checked by observation or experiment, in order to assess the correspondence of the developed ideas to the facts. A hypothesis is elevated to the rank of a theory if no contradiction is found by scientists around the world, or even if there are contradictions, since it is practically impossible to simultaneously link all available facts, models and theories. As a result of these efforts, a certain scheme (paradigm) is created, which is accepted as a sample of the interpretation of facts. The concept of paradigm means '⋯ recognized scientific achievements that for some time have provided the scientific community with a model for setting problems and solving them' (Kuhn 1969). Science is committed to the construction of axiomatic theories. However, under the pressure of new facts, the axioms are questioned and, in the end, rejected. The paradigm exists until a new hypothesis, under the pressure of new and compelling facts, wins acceptance, which is perceived as a revolution in the relevant branch of science. The style of research is changing; as they say (Kuhn 1969), the paradigm is changing. The transition is very painful, and, as Planck wrote (1936, p 97), 'an important scientific innovation rarely makes its way by gradually winning over and converting its opponents: it rarely happens that Saul becomes Paul. What does happen is that its opponents gradually die out and that the growing generation is familiarized with the idea from the beginning'.

From the point of view of social efficiency, theories that include the largest number of facts are the most useful, and society should support efforts in this direction, first of all. However, there are many random circumstances: to get funding, a group of researchers can support deliberately incorrect theories, the only advantage of which is that they are their own and generate revenue. The strongest ones survive.

12.4.3 Organization of scientific research

New knowledge appears in the minds of individuals. These individuals are united in formal and informal groups, institutes, laboratories ···. The formal organization includes universities, laboratories, and institutes ···. An informal organization is based on accepting the same value system, accepting the same paradigm: there are invisible colleges, scientific schools ···. Since the life span of individual carriers of knowledge is limited, there are public institutions for storing and transmitting scientific knowledge. There is continuity in science, so the emergence of a scientific school out of nowhere is rare.

Millions of researchers take part in the discovering and transfer of scientific knowledge. For example, in year 1975, in the USSR, 1 223 000 people were employed in 1000 research institutions, including approximately 320 000 candidates of science and 32 000 doctors of science.

Results of science—scientific products are new facts, new concepts, new models (new relations between concepts, models), prediction of new substances, organisms, principles of designing new devices and tools ···. In tangible form, the results of science appear in the form of books, articles in scientific journals, and publications on the Internet. According to Kapitza (1974), approximately 10 000 000 articles a month were published in the world in 1975. In the same year, approximately 100 000 scientific journals in all disciplines and approximately 300 review journals were published. The number of journals and review journals is doubling every 15–20 years, so that by now these numbers have increased significantly, and a huge scientific archive has emerged. The huge growth of publications is due to the fact that the success of the researcher is measured by the number of publications, and therefore there are numerous authors' clones of the original work, and it is difficult to do anything about it.

Scientific knowledge is presented as a system of concepts or as a system of models, and knowledge itself needs to be studied. There was the science of science (scientometrica), which, among other things, describes the structure of scientific knowledge (Dobrov 1966, Nalimov and Mulchenko 1969). There is a need to classify scientific knowledge and scientific activity[3]. Classification is conducted on the basis of different approaches, and therefore any classification is conventional. For example, physical science can be classified according to the object of study (physics of solid body, physics of fluids ···), or by research methods (statistical physics, probabilistic processes ···), but in each study, the object and method are presented simultaneously. Nevertheless, classifications are necessary and important, especially in teaching and education that passes knowledge from one generation to the next.

12.4.4 A career in science

Although science appears to be an abstract system, documented in huge archives, the bearers of knowledge are concrete people. There is such a profession—a scientific

[3] One of the attempts was made by the French physicist André-Marie Ampère (1775–1836), who is also known as one of the founders of the science of electromagnetism.

employee, whose duty is to maintain scientific knowledge, develop and transmit it to future generations. However, it is more accurate to say that research is not a profession, but a lifestyle (Batchelor 1981, 1997). A research associate is a specialist in solving riddles. However, in addition, research activity is a way of existence for a scientific employee, and science provides a means of survival for him. This means that science is a business enterprise with all the characteristics of an enterprise: competition, espionage, clan organization, mafia ⋯.

Let us assume that a certain young person has an intention to devote himself to science, enter the temple of science, or make a career in science. He feels that he has a desire to understand the essence of events, to think about the mysteries of the Universe. He realizes that he has an ability to do research work. As a first step, he graduates from the University, this is necessary, although there are rare exceptions: the known Russian scientist Zeldovich, for example, did not have a higher education. After University, a young man comes to work into a laboratory, and he may be a little disappointed to find that, instead of investigating the fundamental problems of science, he has to do some routine work: perform some standard measurements or calculate digitally the usual integral, which has nothing to do with romantic dreams of new discoveries. The young man begins to understand that he has got to a scientific factory with feudal orders[4], he is surrounded by a lot of people who want to convince him to do stupid work (Watson 1993). Not resigned, he begins to think step by step about his own vision of the problem and, at some stage, he discovers that what the scientific community thinks was true is wrong, a new assumption leads to the ordering of more facts ⋯. This is a new result, a new discovery ⋯. The young man, who is turning into a mature person, presents the discovery to the members of the laboratory, who approve of the logic of the reasoning, but some of the fellows do not welcome the new representation: a clash of interests occurs. Nevertheless, encouraged, our hero, who has now become a professional scientist, presents his results to another group of researchers, he goes to a conference to participate in the discussion on the problem and talk about his results. Although it is impossible to report all the details of the study, the logic of reasoning is unshakeable, the advantages of the new approach are obvious, but ⋯. A new disappointment!

No work is recognized as a contribution to science, part of it, until it is widely distributed through available scientific publications and embedded in the minds of researchers. The article with the results of the research is sent to one of the scientific journals. A positive review is already an assessment of your results. You may be asked to change something. Publication in a prestigious magazine is also a guarantee that your work will attract attention. The summary of your article is published in one or more peer-reviewed journals. This, of course, does not mean that everyone will rush to read your article, putting everything aside; it is just one of perhaps

[4] Norbert Wiener (1956) wrote in the Epilog of his memoirs: 'I am happy that I did not have to be one of the cogs of the modern scientific factory for many years and work only on the tasks specified by my superiors. If I had been born in the present age of intellectual feudalism, I would have achieved little. My heart goes out to the young scientists of today, many of whom are destined to serve as intellectual lackeys'.

100 000 000 articles appearing monthly. But there is still some interest in your article, in the old days you would have received a request to send a copy of the article; now electronic copies are available on the Internet.

Your results are discussed by colleagues, and some references to your articles appear in the articles of other researchers. This shows that your research influences the opinion of the scientific community, and the number of references is used as a measure of this influence. This does not mean, of course, that all researchers agree with you. But, in the end, you were lucky: the scientific community appreciated and agreed with your results, although it noticed some shortcomings and contradictions. Your articles are quoted and discussed in review articles. The results appear in monographs and textbooks. Students learn things that were very obscure 10–20 years ago, things that have occupied your mind for so long ···. A lot of articles, a chaos of facts, variants of a hypothesis, and finally everything falls into place. The tension of searching is replaced by the delight of discovery!

But, obviously, the described happy development of the event is not the fate of every article out of the total number of 100 000 000! To be successful, you must deal with a problem, a problem that many people recognize must be solved. The choice of research topic determines the value of the scientist, but, they say, the scientist does not need to choose the problem, the problem itself finds the researcher. Further, the research proceeds in some organizational framework that partly exists in your institute and partly should be created by you. At the same time, your organizational skills and reputation are important[5]. You must show your diplomatic skills so as not to lose your success in attracting honorary authors and by virtue, for example, of the Matthew effect (Merton 1968, 1988).

However, business is business. All means are good to get money to finance the problem. We must somehow exclude competitors, of course, on a principle, scientific basis, and the struggle for money manifests itself in the form of a struggle of ideas in science. A good way to get funding is to invent a proper new and important problem: the Millennium bug, for example, or global warming. A global catastrophe is also good for this purpose. There are also fraud, data fabrication, plagiarism, and other questionable methods in science (Marshall 2000). From time to time they remember about ethical standards in science, but this is no longer the main thing.

References

Altschuler G S 1979 *Tvorchestvo kak tochnaja nauka [Creativity as a Science]* (Moscow: Sovetskoe Radio)

Barenblatt G I 1982 *Podobie, avtomodel'nost', promezhutochnaja asimptotika: Teorija i prilozhenija k geofizicheskoj gidrodinamike, [Similarity, Self-Similarity, Intermediate Asymptotics: Theory and Applications to Geophysical Hydrodynamics]* 2nd edn (Leningrad: Gidrometeoizdat)

[5] As J J Hopfield (1997) remarkably put it: 'Reputation is essential to obtain research support, employment, and promotions, and it determines career trajectories in science. The largest determinant of scientific reputation is the papers that bear one's name. The ability to present insightful seminars, nurture young researchers, and informally exchange useful information also affect reputation.'

Batchelor G K 1997 Research as a lifestyle *Appl. Mech. Rev.* **50** R11–20
Batchelor G K 1981 Preoccupations of a journal editor *J. Fluid Mech.* **106** 1–25
Bekhtereva N P 1999 *Magija mozga i labirinty zhizni [Magic of the Brain and Mazes of Life]* (Saint Petersburg: Notabene Publishing House)
Birkhoff G 1969 Mathematics and psychology *Soc. Ind. Appl. Math.* **11** 429–69
Bychkov V L and Zaytsev F S 2019 *Matematicheskoe modelirovanie jelektromagnitnyh i gravitacionnyh javlenij po metodologii mehaniki sploshnoj sredy [Mathematical Modeling of Electromagnetic and Gravitational Phenomena according to the Methodology of Continuum Mechanics]* 2nd edn revised and expanded (Moscow: MAX Press)
Carlyle T 1837 *The French Revolution: A History* (London: Chapman and Hall)
Chandrasekhar S 1987 *Truth and Beauty–Aesthetics and Motivations in Science* (Chicago, IL: University of Chicago Press)
Craik K J W 1943 *The Nature of Explanation* (Cambridge: Cambridge University Press)
De Bono E 1970 *Lateral Thinking: Creativity Step by Step* (New York: Harper and Row)
Dobrov G M 1966 *Nauka o nauke [Scientometrics]* (Kiev: Naukova Dumka)
Dunne J W 1927 *An Experiment with Time* (London: A & C Black)
Frazer J G 1922 *The Golden Bough: A Study in Magic and Religion* (New York: MacMillan)
Haken H and Haken-Krell M 1992 *Erfolgsgeheimnisse der Wahrnehmung Synergetik als Schltissel zum Gehirn* (Stuttgart: Ullstein Deutsche Verlags-Anstalt)
Hadamard J 1951 *An Essay on the Psychology of Invention in the Mathematical Field* (Princeton, NJ: Princeton University Press)
Hopfield J J 1997 Authorship: Truth in Labeling *Science* **275** 1403
Hopfield J J 1999 Brain, neural networks, and computation *Rev. Mod. Phys.* **71** S431–7
Kapitza P L 1974 Centenary of the 'Journal of Experimental and Theoretical Physics' and the role of journals in the development of science *Sov. Phys. Usp.* **16** 928–33
Karlin S 1959 Mathematical Methods and Theory in Games *Programming, and Economics* (London: Addison-Wesley)
Krasnikov S V 2015 *Nekotorye voprosy prichinnosti v OTO: 'mashiny vremeni' i 'sverhsvetovye peremeshhenija': Osnovnye idei i vazhnejshie rezulataty za poslednie 20 let [Some Questions of Causality in GRT: 'Time Machines' and 'Faster-Than-Light Travel': Key Ideas and Major Results over the Past 20 Years]* (Moscow: LIBROKOM/URSS)
Kuhn T S 1969 *The Structure of Scientific Revolutions* (Cambridge, MA: MIT Press)
Kulakov Y I 2004 *Teorija fizicheskih struktur [Theory of Physical Structures]* (Novosibirsk: Alfa Vista)
Lilly J C 1987 *Programming and Metaprogramming of the Human Biocomputer: Theory and Experiments* (New York: Julian Press)
Mach E 1883 *Mechanik in ihrer Entwickelung historisch-kritisch dargestellt* (Leipzig: F.A. Brockhaus)
Mach E 1906 *Erkenntnis und Irrtum. Skizzen zur Psychologie der Forschung* (Leipzig: J.A. Barth)
Marshall E 2000 How prevalent is fraud? Thatas a million-dollar question *Science* **290** 1668–9
Merton R K 1968 The Matthew effect in science *Science* **159** 56–63
Merton R K 1988 The Matthew effect in science II: cumulative advantage and the symbolism of intellectual property *ISIS* **79** 606–23
Nalimov V V 1989 *Spontannost' soznanija: Verojatnostnaja teorija smyslov i smyslovaja arhitektonika lichnosti [Spontaneity of Consciousness: Probabilistic Theory of Meanings and Semantic Architectonics of Personality]* (Moscow: Prometej)

Nalimov V V and Mulchenko Z M 1969 *Naukometrija: Izuchenie razvitija nauki kak informacionnogo processa [Scientometry: Studying the Development of Science as an Information Process]* (Moscow: Nauka)

Pimenov R I 1991 *Osnovy teorii temporalanogo universuma [Fundamentals of the Theory of the Temporal Universe]* (Syktyvkar: Komi scientific center of UrO of the USSR Academy of Sciences)

Planck M 1936 *The Philosophy of Physics* (London: George Allen & Unwin) Johnston H W

Popper K R 1959 *The Logic of Scientific Discovery* (Abingdon: Routledge)

Popper K R 1963 *Conjectures and Refutations: The Growth of Scientific Knowledge* (London: Routledge and Kegan Paul)

Segeth W 1973 *Elementare Logik* (Berlin: VEB Deutcher Verlag der Wissenschaften)

Simon H A 1962 *The Science of the Artificial* (Chicago, IL: Chicago University Press)

Turchin V F 2000 *Fenomen nauki: Kiberneticheskij podhod k evoljucii [Phenomenon of Science: Cybernetic Approach to Evolution]* 2nd edn (Moscow: ETS)

Robb A A 1936 *Geometry of Time and Space* (Cambridge: Cambridge University Press)

Vladimirov Y S 2009 *Metafizika [Metaphysics]* 2nd edn revised and expanded (Moscow: BINOM Laboratorija znanij)

Vladimirov Y S 2015 *Priroda prostranstva i vremeni: Antologija idej [The Nature of Space and Time: An Anthology of Ideas]* (Moscow: LIBROKOM/URSS)

Watson J 1993 Succeeding in science: some rules of thumb *Science* **261** 1812–3

Wiener N 1956 *I am a Mathematician: The Later Life of a Prodigy* (New York: Doubleday)

Yaglom I M 1980 *Matematicheskie struktury i matematicheskoe modelirovanie [Mathematical Structures and Mathematical Modeling]* (Moscow: Sovetskoe Radio)

www.ingramcontent.com/pod-product-compliance
Ingram Content Group UK Ltd.
Pitfield, Milton Keynes, MK11 3LW, UK
UKHW051854100926
8556IPUK00005B/620

9 780750 334495